Middle East in Focus

Series Editor
Barry Rubin
Rubin Center for Research in International Affairs
Herzliya, Israel

The Middle East has become simultaneously the world's most controversial, crisis-ridden and yet least-understood region. Taking new perspectives on the area which has undergone the most dramatic changes, The Middle East in Focus, edited by Barry Rubin, seeks to bring the best, most accurate expertise to bear for understanding the area's countries, issues, and problems. The resulting books are designed to be balanced, accurate, and comprehensive compendiums of both facts and analysis presented clearly for both experts and the general reader.

More information about this series at
http://www.palgrave.com/gp/series/14804

David Rosenberg

Israel's Technology Economy

Origins and Impact

palgrave
macmillan

David Rosenberg
Efrat, Israel

Middle East in Focus
ISBN 978-3-030-09546-8 ISBN 978-3-319-76654-6 (eBook)
https://doi.org/10.1007/978-3-319-76654-6

Printed on acid-free paper

This Palgrave Macmillan imprint is published by the registered company Springer International Publishing AG part of Springer Nature
The registered company address is: Gewerbestrasse 11, 6330 Cham, Switzerland

To my father

Acknowledgements

Amotz Asa-El, Rob Daniel, Pinchas Landau, Elhanan Miller, David Wexler, Benjamin Wurzburger and Shimshon Young generously took time to review various chapters of this book as it was being written. Any errors or omissions, of course, are my responsibility. In addition, the late Barry Rubin played a critical role in bringing this book to publication and encouraging me to undertake the project.

My wife, Abigail Hazony, allowed me the time to complete the project. Various children and stepchildren did their part by, against all reasonable odds, never disturbing or accidentally destroying any notes or papers or causing any damage to my laptop.

Contents

CHAPTER 1

Introduction

The summer of 2011 saw Israelis pour into the streets in a wave of mass protests and street encampments in an often inchoate but heartfelt call for a society that better served its constituents. The protests came in the wake of the Arab Spring that had erupted across the Middle East in the months before and in many respects foreshadowed the Occupy Wall Street rallies that would begin shortly afterward in the USA and Europe. But the Israeli movement was very much in a sui generis phenomenon. A Facebook protest against the rising price of cottage cheese, a food that is a staple of the Israeli diet and for many is a symbol Israeliness, quickly grew into a consumer boycott. Street protests followed after a 25-year-old film editor pitched a tent in Tel Aviv's Habima Square to protest her inability to find affordable housing in the city. She was joined by others creating a tent city that rapidly spread up and down adjacent Rothschild Boulevard. Next came a series of rallies in Tel Aviv and elsewhere around the country that at their peak in early September drew close to half a million people in a country of about 7.5 million.[1] By then, the grievances had widened to encompass rising home prices and the high cost of living generally, the ineffectiveness of government, and growing inequality, not just in terms of income but in apportioning the burden of taxes and army service.

The character of the protests was something new for Israel. As much as they were subject to debate among the protesters themselves, the grievances principally addressed the concerns of the country's middle class, not of the poor or of the traditional interest groups around which

D. Rosenberg, *Israel's Technology Economy*, Middle East in Focus,
https://doi.org/10.1007/978-3-319-76654-6_1

Israeli politics is usually arranged—ultra-Orthodox Jews (Haredim), West Bank settlers, Jews of Middle East and North African origin, and the country's Arab Palestinian minority. The middle-class character of the protest was symbolized by a series of demonstrations during the summer by young parents with their children dubbed the March of Strollers. A survey by the Israel Democracy Index (IDI), conducted in the wake of the protests, provided quantitative evidence of the protests' middle-class character. An annual measure of public attitudes toward government, society, and current affairs, it found that more than a quarter of all Israelis said they personally participated in the protests during the summer of 2011.[2] That figure, of course, is based a self-reported information and probably overstates the actual level, but given the actual turnout at the protests during that summer there can be little doubt that they drew a large part of the population. The IDI survey found that the crowds who gathered in Tel Aviv and other cities that summer were in the main from Israel's middle- and upper-middle class. It found that among those who said they participated in the protests, the highest rates were those claiming "income slightly above average" (40.0%) and "income well above average" (32.2%).[3] The lowest rate was among those reporting "income well under average" (16.5%).

Another important aspect of the Israeli social justice protests was that in contrast to their counterparts in the Arab world or in the USA and Europe they came amid a period of seeming peace and prosperity for the Jewish state. The last major spasm of violence that Israel had been forced to contend with was Second Intifada, which claimed more than 1000 Israeli and 5550 Palestinian lives, but by 2004–2005, it had wound down. In the following six years, Israel fought the 2006 Lebanon War and the 2008–2009 Operation Cast Lead offensive against the Palestinian-ruled Gaza Strip. But by Israeli standards, they were of no consequence—short conflicts that had no long-term economic effect and little psychological impact. Indeed, few if any of the voices of that summer were calling for Israel to reach an agreement with the Palestinians on the assumption that Israel's perpetual state of war was an unacceptable economic burden that peace could solve. The economy had been enjoying unbroken growth since 2003, including the worst years of the global recession in 2008 and 2009. During the majority of those years, economic growth exceeded 4% annually and came close to 6% in two of them.[4] As the social protests were erupting in the third quarter of 2011, the unemployment rate had fallen to 5.5%, close to its lowest level in decades. Over the two decades prior to the 2011 social justice protests, Israel's per capita income climbed into the ranks of the world's richest countries to reach $31,470, close to the average for European Union countries.[5] Life expectancy on average for Israelis was about 81.5 years in 2009, among the highest in the OECD.[6] During the years 2003–2012, the IDI's annual poll asking Israelis to assess the country's "general situation" saw the percentage responding "very good" or "pretty good" rise from 11.1 to 38.1% (although it should be noted that those answering "so so" remained the single largest category).[7]

In most respects, nothing has changed since 2011: The economy has continued to show strong top line economic growth and has demonstrated an enormous capacity for creating jobs. Real GDP growth averaged 3.8% annually in the years 2003 through 2014, well over twice the average for countries belonging to Organization for Economic Cooperation and Development (OECD).[8] The unemployment rate was 5.3% on average in 2015, its lowest in three decades, even as more people entered the workforce and was much lower than the OECD average of 7.9%.[9] Long-term unemployment was 1.9%, the third lowest on the OECD. All of this marked a signal achievement for a country that has few natural resources, bears heavy defense costs, and lies in a regional of perpetual political instability. Just a generation earlier, few Israelis had any expectations that they would achieve Western levels of prosperity anytime in the foreseeable future.

The social protests faded out in the autumn of 2011 and efforts to revive them the following spring and summer failed. But that should not detract from their significance because they represented an economic and social angst that justifiably remains very much present in Israel. On a wide range of social and economic indicators, the country's performance relative to the world's wealthiest economies, which is properly Israel's benchmark, has been poor. On a per capita basis, Israeli economic growth has outpaced OECD countries by a narrower margin of about 1.9 to 1 annually on average during 2003–2014.[10] Israel has narrowed the per capita GDP gap with the wealthiest OECD countries over the decade to 2014 by about a third, but in labor productivity it lags far behind and the gap has changed little, which points to structural problems the economy has yet to solve.[11] At 18.7% of the population, Israel had the second highest income poverty rate among OECD countries in 2013, and was well above the OECD average of about 11%, despite more than a decade of slower economic growth and much higher jobless rates in Europe.[12] Among all Israelis currently employed, some 37%

reported that they found it "difficult" to live on their current income and 12% said they felt poor.[13] Israel society is characterized by unusually wide income inequality: Measuring the ratio between people in the highest and lowest income deciles, Israel is among the most unequal societies in the OECD. Only Mexico and the USA showed wider gaps.[14] While Israelis suffer unusually high housing costs, they also suffer more crowdedness (1.16 rooms per person on average versus 1.7 for all the OECD).[15] The Israeli middle class, as defined by families earning between 75 and 125% of the country's median income, had been shrinking and on the eve of the social protests constituted barely half the population.[16] The cost of living for Israeli families, as a spate of media reports showed during and after the protests, is high relative to Western Europe and the USA, with identical products costing more for the Israeli consumer than his American or British counterpart despite his lower spending power.[17]

Israeli schools have failed to deliver an education commensurate with the needs of an economy whose main resource is its population's intellectual capital. In the OECD's PISA test, which is used to evaluate national education systems around the world, Israelis routinely score at the bottom of the world's developed economies despite their country's obvious successes in science and technology, such as patents per capita and global rankings of research universities. In the 2012 test, the average student in Israel scored 474 in reading literacy, math, and sciences, versus the OECD average of 497.[18] Israel's high levels of income inequality percolate down into the schools, where lower socioeconomic status translates into some of the widest disparities in PISA math scores among OECD countries.[19] Indeed, Israeli education seems to be characterized by a reliance on the personal initiative and skills of its best students and teachers from pre-school through the universities to successfully navigate a system that is bureaucratic and inefficient, a situation anecdotally evidenced by the strong performance of the universities in global rankings even as the government has starved them of funds over the past two decades.

One measure of the extent of personal dissatisfaction of Israelis, in particular the absence of economic opportunities, is the extent of emigration. During the long years of economic malaise in the 1970s and 1980s, emigration from Israel was very high. As the economy recovered in the 1990s and into the twenty-first century, the pace slowed, not counting the phenomenon of newly arrived Russian immigrants returning home after a few years in Israel. By one estimate, the number of native-born Israelis living abroad is not exceptionally high relative to other developed countries (5.85% of the population, versus a median of 4.9% for countries belonging to the OECD).[20] But the profile of Israelis living abroad is revealing, especially for an economy where the best and largest number of opportunities are conventionally thought to be in high technology and other knowledge-based industries. In the USA, Israeli immigrants are on the average younger and are more highly educated than the population at home. Indeed, the educational level of Israelis living in the USA surpasses non-Hispanic Americans, and their earnings surpassed their American-born peers a few years after they arrived.[21]

In short, Israel's economy has made remarkable strides in the past two decades, but in many respects it has failed to achieve the economic and social parameters of the world's wealthiest economies.

Some of this can be laid to the fact that Israel faces unusual, if not unique, challenges to economic growth and social development. Among them is the high cost of security and political uncertainty, an inescapable consequence of its regional setting in the Middle East. In the absence of any conventional powers on its borders that have the ability and/or will to fight Israel, Israel's existence is less threatened than any time in its history. But the array of unconventional forces in Lebanon, the Gaza Strip and Syria with the rise of Islamic movements, Israel is faced with repeated micro-security challenges that manifest themselves in the form of short but frequent missile wars. Over and above the security issue, Israel has had to forge a society from successive waves of immigrants, representing a wide range of cultures, education, and social development over its 70 years. That has required a vast investment not only in creating jobs, constructing housing, and developing infrastructure but also in integrating disparate cultures into a single society. On the whole, this process of "immigrant absorption," as it is known, has been successful, but two homegrown social challenges have emerged in the last decade among the growing minorities of ultra-Orthodox (Haredi) Jews and Israeli Arabs. For differing reasons, both groups participate in the labor force at much lower levels than other Israelis and suffer much higher rates of poverty and lower rates of education.[22] Among Haredi men, 53.7% of the adult men were working in 2015,[23] while among Israeli Arab women, the rate was 22% in 2011.[24] By comparison, the rate for other Jewish males in Israel was 90.8% and for Jewish females 80% in those years. Neither minority has been equipped with the skills and education required to work in an advanced economy. The problem is magnified by the inability of Israel to integrate either group into wider society—sharing its values and aspirations—a critical failure in an economy that is so highly reliant on intellectual capital and social solidarity.

Apart from the unique factors of security, immigration and the social integration of Haredim and Arabs, Israel has shared the same challenges other developed economies are contending with as they move away from the economic model of the twentieth century, where the industrialized economies of the West mass-produced goods for a consumers enjoying ever-growing incomes and social equality. Under this system, where market forces failed, governments were ready and willing to step in to ensure minimum standards of living as well as access to health and education. The system began to come apart as early as the 1970s. Competition from Asia destroyed many manufacturing industries and forced others to reduce costs and payrolls to remain competitive. In an effort to restore competitiveness and growth, governments began to pare back their involvement in the business sector, removing or reducing many regulations, privatizing state-owned businesses, and scaling back income-transfer programs that had helped ensure growing income equality. Measured by the Gini coefficient, which gauges equality (using a scale of zero to one, with zero being the most equal), OECD countries saw their average score rise about 10% from 0.29 in the mid-1980s to 0.316 in the late 2000s.[25] Israel's Gini coefficient also rose in those years by an even sharper 12% from 0.33 to 0.37.[26] Unlike much of the West, industry always accounted for a smaller proportion of the economy in Israel than services, and there was no heavy industry to speak of at all. But Israel did have a high degree of unionization in the first three decades of the state and other government measures were taken that ensured a high degree of income equality. But, like elsewhere in the West, Israel's system began to unravel in the 1970s and 1980s, with the decline of old industries such as textiles and its inability to create a globally competitive economy.

In the place of dying industrial economies, policymakers in the West and the more advanced economies of Asia have looked to knowledge industries as a successor model—a means of ensuring sustained economic and productivity growth that, in turn, generates well-paid jobs through the development and creation of high-value-added products and services. All of these are qualities that the developed economies of the West can offer in greater abundance than the rising industrial economies of Asia. The problem for this strategy is that it is not at all clear that knowledge businesses can provide the same scale of employment and income equality as the great manufacturing businesses of the twentieth century. One glaring example is Apple, which in many respects epitomizes a knowledge-based company in the early twenty-first century. Apple generates huge profits, which in 2012 amounted to $400,000 per employee. Those profits were derived mainly from the knowledge and skills of its US workforce, which comprises in the main engineers, designers, and other high-skilled people.[27] Yet Apple in 2012 employed just 43,000 people in the USA (and 20,000 overseas). That was just a fraction of over 400,000 American workers at General Motors in the 1950s or the hundreds of thousands at General Electric in the 1980s. The bulk of employment at Apple has been moved to outside contractors, whose combined payrolls amounted to some 700,000 people who build and assemble its iPads, iPhones and other products. For reasons of cost as well as industrial capacity, almost none of them work in the USA. While those 43,000 American Apple employees work under excellent conditions, as measured by pay, job satisfaction, and other conditions, there were many other Americans who had no access to the value-added created by Apple. In the second decade of the twenty-first century, there is a growing realization that the knowledge economy produces losers as well as winners.

Knowledge economies are those where information and intellectual resources are the primary generators of growth and value-added, coming in place of agricultural and industrial production that had traditionally played that role. The foundation of a knowledge economy is investment in research and development, creation of human capital through education and training, knowledge sharing, and an efficient system of ensuring knowledge rights (patents).[28] Start-up companies are at the apex of the knowledge economy—the institutions that make use all these elements while conversely embodying few of the elements that go into older production-based economies. Together with California's Silicon Valley, Israel emerged in the 1990s as a global center of innovation in computers and communications technology and was an early adopter of the start-up company as business model and catalyst for developing and commercializing new products and services. Thus, Israel's thriving start-up sector would seem to make Israel a knowledge economy of the first order. However, closer examination of the components of a knowledge economy tells a different story. Berglind Asgeirsdottir lays out the four pillars of a knowledge economy[29] against which INSEAD's Global Innovation Index for 2012[30] and some other measures provide a quick, if crude, performance rating for Israel. All of these are examined in greater depth in subsequent chapters. A short summary follows.

The first of the four pillars is innovation, which Asgeirsdottir measures by such factors as levels of research and development spending and patent filings. Israel has certainly demonstrated a strong capacity in all these areas, both in terms of resources directed at innovation, such as R&D spending, and in results, as measured by such parameters as patents and the creation of start-up businesses. Israel ranks No. 1 globally in the INSEAD survey for gross expenditure on R&D (GERD). Israel was No. 7 in the INSEAD rankings of Patent Cooperation Treaty (PCT) applications per billion dollars of gross domestic product.[31] In terms of knowledge diffusion, Israel ranks 12th in the world, according to the INSEAD survey—sixth for computer and communications service exports and 15th for net foreign direct investment outflows. All told, Israel ranks sixth in the world by INSEAD's definition of knowledge creation. Asgeirsdottir notes that innovation has forced faster product cycles, which means that a country's innovative capacity is also gauged by the extent to which its companies seek new ways of acquiring innovation via links to universities, mergers and acquisitions, and/or alliances with each other. In that respect, as well, Israel's high-tech industry—as distinct from the country's other business sectors—has leveraged its knowledge resources quite well. The sector is dominated by small R&D-focused companies, whose technology prowess can be measured in lieu substantial sales by their ability to attract large amounts of cross-border investment from technology multinationals and foreign venture capital. By virtue of its small home market, Israel is by its very nature global, forming cross-border links through strong ties with Silicon Valley and other technology centers and alliances with overseas companies. Its universities and its defense establishment are important sources of innovation. In the 2004–2013 period, between 25 and 110 Israeli high-tech companies were either merged with or acquired annually, nearly all of them by foreign firms, although the fact that these companies were in the main tiny means that the average deal sized ranged from as little as $30 million some years to no more than $120 million in the best years.[32] Its high-tech companies tend to list overseas, mainly on the Nasdaq, rather than on the Tel Aviv Stock Exchange to enhance their "global" credentials from the perspective of investment, customer recognition, and branding. In 2003–2012, Israeli tech companies raised $1.69 billion in 25 initial public offerings in the USA and another $1.16 billion in 23 IPOS in Europe, while raising just $493 million, spread across 55 companies, in Tel Aviv.[33] More than half of all venture capital investment in Israeli start-ups is made by foreign funds and the lion's share of capital deployed by Israeli VCs comes from overseas.[34]

The second pillar is the use of new technologies by business and government, including those outside high-technology industries. By this measure, Israel as a whole has lagged its peers in the West. Companies outside of Israel's high-tech sector tend to be focused on the domestic market, where competition is limited by market dynamics or government regulation. Under the circumstances, there is little incentive to adopt new technology as a means of either increasing productivity or creating a competitive edge. In government, including the state-owned industries that play a major role in the economy, there is similarly little incentive to make use of new technologies. Many of the parameters that are used to illustrate Israel's knowledge-economy credentials, such as R&D spending as a percentage of GDP and knowledge-intensive employment (where Israel ranks 15th) reflect the resources concentrated in a small sector of the economy and the outsized presence of R&D centers operated by multinational companies. In INSEAD measures that seek to capture business sophistication in broader terms, such as trade and competition, Israel ranks No. 40, with a relatively low percentage of the economic output going to imports (ranked 91) and exports (73).[35] Israelis' perception of the intensity of local competition ranked it at 25 in the world in the INSEAD survey. Vis-a-vis government, Israel ranks high in terms of online government services (15) and the use of the Internet to provide government services (7), but the government scores poorly on broader issues, such as the regulatory and business environments it has created (62 and 25, respectively). This does not mean that the rest of Israel's economy is bereft of knowledge assets; indeed, some areas, such as the defense industry, are rich in them. Individual companies and some smaller sectors are as well. But as a rule businesses outside of the high-tech industry rely to varying degrees on other assets, typically access to natural resources, or a monopoly or near-monopoly position in the domestic market, and/or a favorable regulatory environment.

The third pillar Asgeirsdottir talks about is human capital, namely the knowledge, skills, and competences of the working population critical for developing a knowledge economy. She notes that there is an established relationship between human capital and labor productivity, such that the first two pillars of innovation and new technology are not effective without a stock of trained and qualified workers. By INSEAD's measure, Israel ranks fourth in the world in human capital and research, but that reflects its high rankings in the more rarefied segments R&D. In terms of primary and secondary education, Israel scores poorly on public expenditure per pupil as a percentage of GDP per capita (61) and on PISA tests for reading, math, and science (39). For the percentage of the population enrolled in any institute of tertiary education, it ranks 43. On non-education measures, Israel's ranks surprisingly low. Access to information and communications technologies (ICT), as measured by factors such as Internet users and mobile broadband subscriptions per 100 population, Israel ranks in the INSEAD survey 19th. It ranks 21st for ICT access. Not all of Israel's rankings in the INSEAD are terribly poor. In online creativity, for instance, which gauges such informal activity as Wikipedia edits and video uploads on YouTube, Israelis are remarkably active (ranking, respectively, fifth and ninth in the world, respectively, on a per capita basis). Taken against the relatively low ICT usage, it suggests that Israelis' creative abilities are limited to a small but intensively active part of the population, much as in business the intense focus on R&D is concentrated in a single sector of the economy. Overall, for an economy that is more reliant than others on its intellectual capital Israel's ranking in innovation is not especially impressive. As Chapter 7 shows, Israel's schools contribute relatively little to preparing its young for life in a knowledge economy; rather, as Chapter 8 explains, it is cultural and other characteristics that make the country an ideal breeding ground for start-up companies.

Asgeirsdottir's fourth pillar is what she calls enterprise dynamics, which comes principally from newly created firms. Start-ups in technology and other fields are responsible for a disproportionate amount of innovation and account for an increasing share of private sector R&D and patent activity in the USA and some OECD countries. Additionally, she cites social and organizational changes that have accompanied the rise of the knowledge economy, which put a greater emphasis on teamwork and flatter management structures that demand greater initiative and personal responsibility on the part of lower-level employees. Many of these qualities have traditionally existed in Israeli society and became an integral part of business and other organizations. That has fostered Israel's transition to a knowledge economy and its particular emphasis on the start-up company, which embodies all these new business dynamics. But in the enthusiasm to adopt the ideals of a knowledge economy, it can easily be forgotten that large organizations require a degree of discipline, rules-based procedures, and hierarchy to manage a large and complicated array of resources, ranging from employees to capital to production and logistics. In these areas, Israel has been demonstratively less successful to the extent that arguably two economies exist side by side—one that is internationally competitive by employing innovation and technology and a second that is geared to the domestic market and relies on monopolistic markets and regulation. Indeed, the Israeli business environment is difficult for new and small businesses, start-up companies being the prominent exception. The rate of entrepreneurship overall remains low, with just 10% of the adult population saying they were engaged in a newly established enterprise, 36th of 67 countries surveyed.[36]

Asgeirsdottir does not cite it as a pillar of the knowledge economy, but the growth and development of an industrial cluster—a geographic concentration of businesses, suppliers, and associated institutions like finance and universities focused on a particular industry—seems to be as critical factor for high technology as it was in the past for other industries. This would seem counterintuitive in the Internet age, in particular in regard to an industry that itself was built on information and communications technology and by its nature an early adapter and heavy user. But the evidence of the importance of clusters, which brings together academics, entrepreneurs, engineers, and investors, as well as the physical infrastructure of office space, easy transportation, and even cultural and entertainment offerings, is overwhelming.[37] In that respect, Israel has built a hugely successful cluster, as evidenced by its No. 5 ranking in the global Start-Up Ecosystem Ranking for 2015 of 20 centers.[38] Outside the USA, it is the No. 1. The report captures on a micro-level the importance of geography, or more exactly proximity, in start-up culture. "In-person conversations lead to innovation, especially for early-stage start-ups where the strategy is likely to change three times between 9 a.m and 5 p.m, and the best work is often done by a core team after midnight over late-night pizza delivery. Success requires moving fast and pivoting even faster, in a race to find product/market fit before the money runs out. Often there is precious little time to send thoughtful updates to far-flung employees or account for multiple time zones. Look at the office layout of early-stage start-ups and often you won't even find desks separated. Instead, the whole team sits around one large table so they can all hear every conversation and informally stay on the same—fast moving—page."[39]

The Israeli start-up phenomenon is a consequence of two factors. The first is the country's intellectual capital. Long before the country's high-technology industry emerged, Israel and before that the pre-state Jewish community, commonly known as the yishuv, had created the basis for a knowledge economy. Universities and other institutions of higher education were established and prospered in a poor and underdeveloped economy that struggled to find an outlet for their graduates' training and skills. Many of the immigrants who arrived starting at the turn of the century were well educated by the standards of the day. That phenomenon was reinforced in the 1930s, with the arrival of German immigrants escaping Nazism, and against in the 1970s and 1990s with two waves of immigrants from the Soviet Union. Their skills were first put to work by the British, who were then ruling what was Palestine, during the Second World War and again in the 1960s and 1970s by the domestic defense industry, which focused its efforts on electronics and communications. By the time the 1990s telecommunications revolution arrived, Israel had the human capital to exploit it.

The second, and arguably the more important of the two components, is the role of a special breed of entrepreneurialism that has developed in Israel. It is characterized by the same culture of risk-taking seen in Silicon Valley and other technology clusters, indeed anywhere where a new and untried industry is emerging. But Israeli entrepreneurialism in technology is also animated by a strong commitment to teamwork and a culture critical of and resistant to rules, conventions and hierarchy. For the majority of Israelis who have served, the army provides an early and formative experience in identifying problems and solving them, not just for those serving in elite technology units but in combat units as well. Measured by the usual criteria of formal educational achievements, Israel's human capital is outstanding, but not superior to countries that have not succeeded in creating a start-up culture on the same scale and intensity. Instead, it is Israeli society's entrepreneurial qualities that have leveraged its intellectual capital into innovative technology. What is remarkable about the concentration of intellectual capital and entrepreneurism in Israel is how culture-specific it is. Unlike California's Silicon Valley and other leading global technology clusters, the Israeli industry has no sizable number of foreign entrepreneurs or engineers. Israel does not serve as a magnet for entrepreneurs and engineers from around the world; rather, its success is very much reliant on a unique cultural brew generated from within Israeli society. Israel's high-technology industry is entirely Israeli as discussed in Chapter 8.

But these human capital resources have not been applied across the entire economy, which is, in fact, two separate economies—one of start-up companies very much engaged in the global economy and competing successfully in it, as measured by the industry's ability to attract cross-border capital and integrate itself into the global supply chain primarily as a source of research, development, and innovation. The other, which accounts for a much larger part of Israeli output and employment, comprises industries geared toward the domestic market, characterized by a lack of competition, low levels of innovation, and productivity as discussed in Chapter 7.[40] Of course, there are segments of the Israeli business sector that belong to neither of these economies—big, export-oriented companies whose business is outside the information technology and communications industries that are outside the core of Israeli high tech. But these segments, while sometimes contributing considerably to the overall economy, do not typify it. They are more likely to be the product of one person's vision and abilities rather than the natural outgrowth of Israel's social, economic, and regulatory environment.

Nor have these human capital resources been applied effectively in Israel's high-tech sector. Israel's industry is routinely compared to California's Silicon Valley, but the human and business geography of California's technology cluster are very different than Israel's in one important respect. Both spawn new enterprises with large stocks of venture capital and supply of ready, experienced entrepreneurs to exploit it. But Silicon Valley has not just generated a stream of start-up companies and innovative, sometimes revolutionary new products and services, it has also created most of the biggest and most important companies in the industry. Not just flagships like Apple, Google and Facebook, Amazon and Hewlett Packard, but hundreds of others medium-sized to large multinationals that are the dominant players in their particular market. They employ not only young engineers and entrepreneurial CEOs but many more people who are required to demonstrate less personal vision, insight, analysis, and endeavor but nevertheless contribute to the enterprise and play a critical role in the industry. These people fill jobs like sales engineers, marketing executives, finance professionals, and a host of others that keep a large enterprise running. In Israel, two decades since high technology has emerged as a distinct industry, companies of this scale and range are few and far between. Israelis acknowledge this phenomenon and the most optimistic among them point to signs that the industry is maturing and that larger enterprises with fundamental technology, experienced management and long-term business strategies will emerge. However, that transition will not occur so easily, if at all, because the very same characteristics that have enabled the rise of Israel's start-up culture, as described in Chapter 8, perversely prevent it from fully exploiting its innovative abilities.

Israel has become a knowledge economy over the past two decades only in the narrow sense that it has developed a globally competitive start-up sector focused principally on the development of information technology and, more recently, life science companies. But most of the economy has not been part of this process. Moreoever, Israel's start-up sector is suffering the Apple phenomenon, albeit on a much smaller scale, in its inability to generate enough high-quality jobs to ensure improving levels of social and economic equality whose impact is felt through the entire economy. This book seeks to chart the history, structure, and factors behind this phenomenon and to explore its outcomes.

Notes

1. Oz Rosenberg, "Some 450,000 Israelis March at Massive 'March of The Million' Rallies Across Country," *Ha'aretz*, September 3, 2011.
2. Tamar Hermann, *The Israel Democracy Index 2012*, 121.
3. Ibid., 122.
4. Organization for Economic Cooperation and Development, *OECD Economic Surveys Israel 2016* (Paris, January 2016), 7.
5. International Monetary Fund, World Economic Outlook Database, October 2012 edition.
6. OECD, *How's Life? 2015: Measuring Well-Being* (Paris, October 13, 2015), 72.
7. Hermann, *The Israel Democracy Index 2012* (Jerusalem, 2012), 25 and 27.
8. OECD, *OECD Economic Surveys Israel 2016, 7.*
9. Finance Ministry, *Weekly Economic Survey*, August 21, 2016, 9.
10. OECD, *OECD Economic Surveys Israel 2016, 7.*
11. Ibid., 2. The gap between Israel and countries in the upper half of the OECD as measured by GDP per capita on a purchasing power parity

basis was 31.9% in 2014, down from 41.3% in 2003. In terms of GDP per hour worked, the gap was 41.4% in 2014, little changed since 2008.

12. Ibid., 2.
13. Central Bureau of Statistics, Selected Data from the 2013 Social Survey about Employment, 6.
14. OECD, *How's Life? 2015*, 61.
15. Ibid., 69. The inter-decile income ratio for Israel in 2013 was 14.9 versus 12.4 for the OECD 30 on average. In fact, most OECD members have ratios in the single digits; the average is raised by the weighting for population.
16. Bank of Israel, *Annual Report 2011* (Jerusalem, March 28, 2012), 343–344.
17. A Bank of Israel study (*Recent Economic Developments*, No. 132, September–December 2011, 33–38) concluded at the time that prices for food, culture, and leisure services and communications, among others, in Israel were, in fact, high relative to its GDP per capita but that for other consumer goods, such as apparel and footwear, were lower.
18. OECD, *How's Life? 2015*, 79.
19. OECD, *PISA 2012 Results: Excellence Through Equity* (Paris, 2013), 15. The rankings were done by measuring the score-point difference in mathematics associated with a one-unit increase in the PISA index of economic, social, and cultural status. Israel's score difference was 51, the sixth-highest among OECD countries.
20. Yinon, Cohen, "Israeli-born Emigrants: Size, Destinations and Selectivity," *International Journal of Comparative Sociology*, Vol. 52, No. 1–2 (2011), 45–62.
21. Ibid. Cohen cites Israeli and US census data from the 1980s, 1990s, and 2000 as well as studies by himself (1996), Cohen and Haberfeld (2001, 2003), Lev Ari (2008), and Rebhun and Lev Ari (forthcoming). Cohen says there are no studies which estimate the selectivity and economic assimilation of Israelis in European countries or in Oceania.
22. Meirav Arlosoroff and Sivan Klingbail, "Litzman: It's Absurd for Both Parents to Work – It's Enough That the Wife Works" (Hebrew), *TheMarker*, October 31, 2012.
23. Knesset Research and Information Center, *Data on Haredi Employment* (Hebrew) (Jerusalem, March 14, 2016), 7–8.
24. Eran Yashiv and Nitsa Kasir (Kaliner), *The Labor Market for Israeli Arabs* (Tel Aviv, January 2014), 20–21.
25. OECD, *Divided We Stand: Why Inequality Keeps Rising* (Paris, December 2011), 22.
26. Ibid., 23.

27. Charles Duhigg and Keith Bradsher, "How the U.S. Lost Out on iPhone Work," *The New York Times*, January 21, 2012.
28. Dominqiue Foray, "Organizing the Use of Knowledge," in Brian Kahin and Dominique Foray, eds., *Advancing Knowledge and the Knowledge Economy*, 9.
29. Berglind Asgeirsdottir, "OECD Work on Knowledge and the Knowledge Economy," in *Advancing Knowledge and the Knowledge Economy*, 19–21.
30. Soumitra Dutta and Beñat Bilbao-Osorio, eds., *The Global Innovation Index 2012: Strong Innovations Linkages for Global Growth* (Geneva, 2012), 237. All subsequent references to this survey come from this page.
31. PCT refers to a 1970 treaty that provides a single procedure for filing patent applications in all the countries party to the agreement. INSEAD measures that against GDP adjusted for purchasing power parity.
32. Dan Yachin and Oren Raviv, *IATI Annual Review: Israel ICT Industry 2015* (Tel Aviv, January 2015), 5–6.
33. Israel Venture Capital Research Center, *IVC 2013 Yearbook*, 36–37.
34. Yachin and Raviv, 9.
35. Israel's political environment ranking is 64, but the score is misleading. On political stability, Israel is ranked 129, which reflects Middle East regional instability rather than domestic factors and on press freedom, which is based on a Reporters Without Borders annual survey (other world ranking on press freedom score Israel much higher). On government effectiveness, it ranks a relatively high 23.
36. Ehud Menipaz, et al., *GEM Israel 2013 National Summary* (Beer Sheva, Israel, undated), 9.
37. To cite some examples: Annalee Saxenian, *Regional Advantage: Culture and Competition in Silicon Valley and Route 128*; Chong-Moon Lee et al., *The Silicon Valley Edge: A Habitat for Innovation and Entrepreneurship*; and Martin Kenney, *Understanding Silicon Valley.*
38. *Start-Up Ecosystem Report 2015* (San Francisco, 2015), 24. The rankings are based on the performance on the funding and exit valuations of start-ups, venture capital investment and the time it takes to raise capital, the quality of technical talent, its availability and cost, the size of the cluster's GDP and the ease of reaching customers in international markets, and experience of entrepreneurs.
39. Ibid., 17.
40. Bank of Israel, "Labor Productivity in Domestic-Market Oriented and Export Industries: An Analysis from an International Perspective," *Annual Report 2013* (Jerusalem: Bank of Israel), 65.

CHAPTER 2

Origins

Israel's economy and society more than most, maybe all, others is very much an act of will by those who began the work of creating it more than a century ago. The land to which the earliest Zionist pioneers aspired offered little in the way of natural resources on which to develop either agriculture or industry. The traditional economy of the indigenous Palestinians served as neither a foundation nor a model for the European Jews who began arriving in the final two decades of the nineteenth century. The Jewish settlers brought attitudes and ideas with them whose aim was to break free of the structures that had characterized Jewish economic life in their homelands. In effect, they turned the Marxism that many of them espoused on its head: The institutions and relationships they created didn't give rise to an ideology, rather their evolving ideology molded a society created by force of their will.

Many of the hallmarks of those early years, which stretched from the 1880s into the first decades of the Israeli state in the 1940s and 1950s, have little bearing on the knowledge economy that subsequently emerges, but others very much do so. The early Zionists believed strongly in the value of manual labor, of returning to the land as agriculturalists, and in socialism, little of which would be relevant a century later to Israel's high-tech economy. But they also evolved a deeper ethos that proved to be a powerful shaper of Israeli society's capacity for innovation. It was characterized by an enormous capacity for invention and re-invention; self-criticism; egalitarianism; an unusual and often conflicting dual loyalty to the self and to the group; resistance to the authority

D. Rosenberg, *Israel's Technology Economy*, Middle East in Focus,
https://doi.org/10.1007/978-3-319-76654-6_2

of people, ideas, and social conventions; and a sense of national purpose. The diaspora from which they sprung had created a culture of organizing on the fly because the Jews in exile were perpetually in a situation of chaos and uncertainty. Jews and now Israelis created a culture of implementation, which would be utilized with enormous effectiveness as they confronted the multiple tasks of developing an economy and political institutions while reluctantly engaging in a war with the Palestinians. Those values, even if they expressed themselves in very different ways a century ago, would come to the fore at the end of the twentieth century as Israel's start-up industry came into its own.

At the same time, there was a powerful counter-trend, which is less storied in Zionist history and remained in the background well into the 1980s, which viewed the Zionist enterprise in very different terms. It had no pretensions of trying to remake either individuals or society. Rather it saw the Jewish state serving as a shelter for diaspora Jews facing violence and oppression. Jewish society would not be a revolutionary; it would seek to mimic the best of modern Western institutions, including capitalism and the development of expertise and technical skills to develop the impending state along business and scientific lines. As an ideology, this trend of Zionism failed to capture the imagination of the movement's leadership in Palestine. Nevertheless, it became a major element in the life of the *yishuv*, the pre-state Jewish community of Palestine, because so many of the immigrants entering the country, especially those coming in the 1920s and 1930s, tacitly espoused it, even if they failed to create a competitive ideology or movement to advance it.

Although it would take a better part of a century for Israel's knowledge economy to emerge along its current parameters, its deepest origins lay in the land itself. It had virtually none of the natural resources that traditionally underpin modern agriculture and industry. By the final quarter of the nineteenth century, when the first European Jewish settlers arrived, much of the country had been abandoned over the centuries to swamps and desert. The few Jews already living there depended on financial assistance from abroad (*chaluka*). The mainstay of economic life for the majority of Arab population was subsistence agriculture. The Jews of Central and Eastern Europe, who would constitute the great majority of immigrants in the era before Israel was created in 1948, had good reasons for emigrating. Apart from growing anti-Semitism in the Russian empire, the Pale of Settlement, where the majority of Russian Jews were confined, was under immense economic pressure. The industrialization and urbanization that had reached Eastern Europe in the last quarter of the century was threatening the livelihoods of small-time Jewish artisans and merchants at a time when the Jewish population was rapidly growing.

But by the logic of most great waves of immigration, or colonial settlement, the Jews of Eastern Europe had little reason to go to Palestine. Ruled by an Ottoman Empire in the final throes of its decline and fall, it offered neither economic opportunity nor much relief from political and religious oppression. While the British who took control of the area following after the First World War brought improvements in infrastructure and governance, Palestine remained poor and had few economic prospects. In the decades before the outbreak of World War II in 1939, some four million Jews emigrated from Eastern Europe, but only about 4% chose to go to Palestine.[1] The vast majority opted for the USA, Western Europe, Argentina, South Africa, and Australia. The small minority that settled in Palestine was principally motivated by ideology or by the absence of any other choice when their gates to preferred destinations were closed after World War I.

The two ideological trends were in conflict with the establishment of the earliest agricultural communities of the First Aliyah during the last two decades of the nineteenth century.[2] These settlers, known collectively as the Bilu (a Hebrew acronym from the Book of Isaiah for the verse *Beit Ya'akov lekhu v'nelkha*, "House of Jacob, let us go [up]"), adopted an early form of the ideology that sought to turn the Jews into agriculturists working their own land. In fact, they had neither the skills nor the financial resources to succeed, and their settlements faced financial collapse not long after the first of them were established in 1882. Baron Edmond de Rothschild agreed to provide financial and managerial aid that enabled the settlements to prosper but at the cost of jettisoning their ideology to run their farms along business lines. That included the widespread use of indigenous Palestinian labor and the unwelcome supervision by European experts brought in by Rothschild. Under the new regime, one settlement, Zichron Ya'acov, had some 200 Jewish farmers employing 1200 Palestinians.[3] Nevertheless, the socialist trend was far from vanquished. The Second and Third Aliyot (1904–1914 and 1919–1923, respectively) not only brought a new wave of Jews to Palestine but one animated by the political and economic thinking current in Europe, particularly, Russia at the time. In hindsight, much of it seems irrelevant, perhaps quaint, but it created the institutions and ideological guidelines at the heart of the pre-state yishuv and the early years of the Israeli state. If their economics have little relevance to Israel's modern knowledge society many of their personal and social principles surviving into the twenty-first century most certainly do and so they are worthwhile examining.

Those who arrived in the Second and Third Aliyot were believers first and foremost in socialism, some more influenced by Marxist orthodoxy than others but all sharing the view that the society they intended to create would have neither a class of exploiters nor of the exploited. The problem for them was that the social and economic conditions in Palestine did not provide a natural place for people who sought to become the peasants and workers: There was no industry to absorb them, and as farm workers, they were uncompetitive with the cheaper, better skilled and more docile labor offered by local Palestinians. Their second major principle was the value of manual labor, which they viewed not as an economic necessity foisted on them by a primitive economy but as a source of spiritual and social uplift. They saw Jewish existence in Eastern Europe as distorted by anti-Semitic legislation and attitudes that had created a society of petty merchants and other occupations at the margins of the economy who produced little economic value. In Palestine, manual labor and building the land would be the basis of a new Jewish society. "The necessary condition for the realization of Zionism is the conquest of all occupations in the country by Jewish labor" was the slogan appearing in the party newspaper of *Hapoel Hatza'ir*, one of the two main Second Aliyah groups.[4] An important subset of the redemptive power of work was their emphasis of agriculture, an astonishing backward looking ideology given that it was an era when it was evident that industry was the basis for the modern economies in the West and was the chief concern of Marx himself. On the other hand, other values they held would have an important impact on the character of Israeli society in the following decades, among them personal austerity, a strong egalitarianism, and a culture of argumentation and resistance to authority.[5] All of these, it seems, thwarted the rise of the authoritarianism that typically characterizes revolutionary movements and would later contribute to many of the fundamental values of Israel's start-up culture.

While the value of manual labor remained, it was accompanied by a growing realization that education and expertise were critical as well. In fact, both came naturally to the men and women of the Second and Third Aliyot who, as much as they sought to become farm and factory workers, were not predisposed to manual labor by origin or attitude. In the Second Aliyah, which would make up the kernel of Zionist leadership, they were predominantly lower middle class; far fewer were from the poorest classes that made up the majority of Eastern European Jewry.[6] By the standards of the day, they were well educated, with about a quarter having undergone some kind of secondary education and 8% having studied in some framework of higher education. Although their allegiance to the principle of manual labor was certainly heartfelt, many of the leaders of the Second Aliyah in fact engaged in cultural and intellectual pursuits at variance with their own self-image. After several years working in agricultural or road-building, many went on to get university degrees. Among them David Ben-Gurion, the founder of the state, and Yitzhak Ben-Zvi, its second president, both acquired law degrees at the University of Constantinople. Shlomo Zemach, the founder of the Kadoor Agricultural School, studied literature, philosophy, and agriculture in France, and Zalmar Shazar, the third president of Israel, studied philosophy and history in Germany.[7] In that respect, their values merged with those of the second great trend of twentieth-century Zionism, which was directed toward creating a class of experts and technocrats.

The initial answer of the early Zionists to the absence of industry and a capitalist class was to engage in class warfare with the smallholders of the First Aliyah and await the day when enough capital would accumulate to create a capitalist class they could oppose. To their credit, they realized quickly enough that the Marxist models they had arrived with had little application in Palestine and set about to create a series of institutions that were both innovative in their conception and remarkable for their staying power. Foremost among these was the kibbutz, perhaps history's only experiment in agricultural collectivism to have survived more than a generation as a voluntary, democratic, and productive institution. The progenitor of the kibbutz, the *kvutza*, was established by the Second Aliyah at Degania in 1910, but the main characteristics of the kibbutz were created by the Third Aliyah, which did away with family life as well as private property and greatly expanded the size of collective farms from a dozen or so to the hundreds. The era also saw the establishment of many of the Israeli economy's founding institutions—the Histadrut labor federation—and an array of social, health, and education networks as well as businesses and even a bank affiliated with it. As the kibbutz did for agriculture, the Histadrut institutions provided a framework for a working class to find salaried employment in the absence of an employer class. Thus, the Histadrut encompassed both the institutions of labor and capital in a single structure. More than that, it embodied the ideology of the Second and Third Aliyot that economic life was inseparable from political, social, and cultural life.

Although the socialist stream of Zionism prevailed, it had its competitors both ideologically and on the ground in the form of a parallel economy that began to grow alongside it starting in the 1920s. It began with a debate within the movement over competing strategies for settling the Jewish homeland—capitalism versus socialism—that came to a head at the Zionist Congress of 1920. On the one side were American Zionists led by Louis Brandeis, who wanted private investors to spearhead economic development. They argued that the Zionist movement's economic arms, such as the Jewish National Fund, should limit its activities to those of government by building infrastructure and ensuring social welfare, but leave economic development to the private sector. Opposing them was Chaim Weizmann, the champion of European Zionists, who argued that conditions in Palestine were too primitive for capitalism to perform those functions and advocated that the movement buy land and sponsor agricultural settlements. More than that, European Zionists took the view that Zionism wasn't a business enterprise but a social revolution that shouldn't measure itself by return on investment or other capitalist yardsticks. In all events, the capitalist stream of Zionism was strengthened by the next two waves of immigration, which were driven less by ideology than by necessity. The Fourth Aliyah comprised owners of small businesses escaping growing anti-Semitism in Poland in the early 1920s. They spurred the growth of the first cities and buttressed private sector, albeit with small and typically undercapitalized businesses. They were followed in the next decade by German Jews fleeing the Nazis (the Fifth Aliyah). Coming from one of the most industrially advanced economies of the time, they brought greater capital and skills that they used to set up small factories in the cities. Indeed in the years 1932–1937, when the Zionist institutions set up to realize the socialist society were suffering from underfunding, private capital accounted for 87% of all investment in Jewish Palestine.[8] The stress on agriculture by the official institutions of the Yishuv meant that industry was largely left to the private sector to develop.

Finally, World War II would provide a critical boost to private industry: British forces in the Middle East, isolated from their regular suppliers back at home, were forced to rely on local factories and workshops in the machine tool, chemical, textile, steel, and pharmaceutical industries. Shipyards for minesweepers and drydock services were built at Haifa as were landmines and replacement parts for RAF warplanes.

By 1945, about 41% of output of the yishuv's economy derived from industry, compared with 26% in 1936. Only 20% came from agriculture.[9]

An important factor in the rise of the Israeli knowledge economy was the unusually early and prominent role of universities in the Zionist enterprise even at a time when the Palestinian economy had little capacity for making use of trained graduates and the Jewish population it was meant to serve numbered in the tens of thousands.[10] The first proposal to establish an institution of higher education was raised at the First Zionist Congress in 1901 at a time when the Jewish population in Palestine numbered just 50,000. As outlined in the pamphlet, *Eine Jüdische Hochschul*, written by Martin Buber, Chaim Weizmann, and Berthold Feiwel, the university would be dedicated principally to scientific and technological education and research with the aim of stimulating economic development.[11] Their model was the German university system that had performed much the same role in developing the country's industry. But rather than serving as centers for training the next generation of students to take their roles in a mature economy, the purpose of the yishuv's institutes of higher education would be to advance economic and social development.

Remarkably, within 25 years of the pamphlet's publication, two universities had been established—The Technion in 1924, which was wholly devoted to science and engineering, and a year later The Hebrew University, which was devoted to humanities but also had a big component of sciences and medicine. Weizmann himself, dissatisfied with the level of science being done at Hebrew University, helped form a third institution, the Daniel Sieff Institute of Research (later the Weizmann Institute of Science) in 1934. All three institutions were wholly or principally devoted to research and to solving local problems. For instance, the Hebrew University's Geology Department was extensively engaged in searching for underground water resources in the arid countryside while other research focused on intensive irrigation.[12] The immigrants of the Fourth and Fifth Aliyot, sharing little of the infatuation of their predecessors for manual labor, sent their sons to The Technion, so that by 1940 it had produced 1000 graduate engineers, scientists, and skilled technicians, who were employed in the war effort in the years that followed.[13] The universities were similarly engaged in defense research. Hebrew University's Physics Department produced quartz plates used in tanks, planes, and radio transmitters, repaired vacuum tubes, and made items as aerometers, mercury switches, and measurement devices for the

Royal Engineers. At the Technion, laboratories made gasmasks and medicines, and repaired electrical equipment. Its researchers developed new processes for making cartridges, optical instruments, transformers, and communications equipment.[14]

When the British Mandate came to a close in 1948, the yishuv had the basis for a modern economy. It certainly could not compare to the leading industrial powers of the time in either size or sophistication but, thanks to immigration, its local universities and the war effort, it enjoyed an unusually high level of human capital. The War of Independence that established the State of Israel shattered much of the infrastructure that had been developed, but the human resources remained intact and would be enhanced by waves of immigration and the growth of higher education in the decades that followed. However, before it could make full use of those human resources, the nascent state still had considerable challenges to overcome. The state that emerged was still at war with its neighbors, despite truces ending the hostilities in 1949, saddling it with heavy defense costs and political uncertainty. In the years 1948–1952, Israel welcomed into its borders some 690,000 immigrants, almost entirely from the Middle East and North Africa, nearly doubling the population. The immigrants created a massive burden on the economy by requiring housing, infrastructure, and jobs. More than other waves of immigration, it comprised people with little education and few skills, which meant that the return on the investment in absorbing them would be low and long in coming.

In spite of the trauma of war and mass immigration, the yishuv leadership, its institutions, and its ideology made a smooth transition to statehood. The revolutionary socialism of the earlier pioneers had long given way to a more bureaucratic sort that emphasized nation-building rather than class conflict. In 1948, the private sector accounted for about 60% of total production and employed 60% of the workforce.[15] State control over the economy grew in the first two decades, but less by plan or ideological imperative and more by chance or necessity. The government ended up taking control of some of the country's major industries, including the plants that were eventually merged into Israel Chemicals as well as Oil Refineries Ltd. because their private-sector owners were unable or unwilling to maintain them. The flow of capital from the USA and later from West Germany in the form of Holocaust reparations was channeled through the government into state-owned enterprises and infrastructure projects. Side by side, new industries were established by the state and by the Histadrut to encompass defense, public transportation, broadcast media, agriculture marketing and exports, and telecommunications. The government still preferred agriculture to industry (56% of the development budget for 1952–1953 was for agriculture and water and just 11% for industry.)[16] It only abandoned that policy in the middle 1950s when it became too evident that farming could not provide enough jobs to employ the immigrants who had streamed into the country earlier in the decade.

The government's preponderant role did not make for an efficient economy, a problem that was compounded by an ideology that looked askance not only at profit but also at professional management as a means of managing the economy.[17] Nevertheless, economic growth was rapid in the 1950s and first half of the 1960s, thanks to immigration, high levels of (state-directed) capital investment, and an abundance of infrastructure and housing challenges that the government could address with relative efficiency. In the years 1950–1965, gross domestic product grew an average of 10.6% annually and 5.5% on a per capita basis.

For the future knowledge economy, the 1950s and 1960s saw a rapid expansion of the university system that kept intact their basic philosophy and function as service providers to state and society through research and the supply of technologically proficient graduates. The new state needed civil servants, teachers, social workers, economists, and administrators in bigger numbers than ever to manage a burgeoning population and take over the jobs once held by British civil servants.[18] The Hebrew University, by far the most important academic institution, responded by inaugurating undergraduate education in 1950 as well as faculties of law and expanding its offerings in medicine. New universities in Tel Aviv, Haifa, and Beersheva greatly expanded the student population. In the 1990s, academic colleges were established as Israel's population swelled with the arrivals of hundreds of thousands of well-educated immigrants with similar aspirations for their offspring (see below).Over the course of the decade, the student population grew from approximately 76,000 in the 1989/1990 academic year to 166,000 in 1999/2000.[19]

An interesting by-product of the wave of immigration from Middle East and North African countries was the widespread diffusion of the ideology of technology, science, and expertise among the nascent state of Israel's veteran European-origin population.[20] The pre-state ideology that viewed dedication to the cause and selfless labor as more important than expertise began to give way. In its place was something that was less an ideology and more of a strategy that still saw its aims as building the state but redefined the necessary personal and social qualities as skills in management, technology, and science. In the broader economy, oddly, those values would impinge little on the collectivist ethos from the early days—the loyalty to the group that remains a significant factor in Israel's contemporary start-up culture. In a 1965 article in the *Ha'aretz* daily, Shabtai Tevet asks "What happened to the Palmach Youth?", a reference to the elite pre-state fighting force. The answer, he said, is that the old pioneering ethos that expressed itself in politics and the military was now being applied to the private sector, capitalism, and professional management.[21]

Practically speaking this expressed itself in a growing role for the government as the funder of higher education, financing the construction of campuses, and subsidizing tuition. Secondary education was similarly expanded to provide a population of graduates who could go on to university study. That enabled the university population to expand rapidly from 1027 in the 1946/1947 academic year to 6277 in 1959/1960.[22] The strength of Israeli academic research, however, was not matched in business and industry. In 1965, for instance, research and development spending accounted for 1% of GDP, lower than any of the world's industrialized economies, except Italy.[23] Israel counted 10 scientists and engineers for every 10,0000 employees, which wasn't low by global standards, but well below the rate of the USA (25) or Sweden (22).[24] Some efforts at creating civilian technology as early as the 1960s, with Discount Investments and the Elron Group forming companies such as Elscint, which was an early entrant into the medical imaging field as well as the first Israeli company to float shares for trading on the Nasdaq. Until the 1970s, the Industry and Trade Ministry's chief scientist, whose office would later become a major funder of research and development for the high-tech sector, was a part-time position.[25]

It was military technology that would serve as the direct precursor of the knowledge economy, the first industry where innovative technology would be developed in Israel on a large scale. The industry itself dates back to the 1950s, with the establishment of Israel Aircraft Industries (1953) and Rafael (1952), which operated more like a research institute than a company in its first decades and developed some of Israel's first computers. They were later joined by private-sector companies such as Elbit Systems (1966) and Tadiran (1962), which were formed against the background of a dispute over whether to invest in developing and manufacturing platforms and systems at home or whether to rely as much as possible as foreign suppliers. In fact, the debate was largely settled by Israel's inability to source weaponry overseas on a consistent basis, which meant that at least a certain portion had to be produced domestically.[26] All of these developments were given huge impetus by France's arms embargo on Israel shortly after the 1967 Six-Day War. At the time, France was the chief supplier of weapons to Israel, most particularly fighter jets, and the decision by Charles De Gaulle had a profound effect on the country's defense posture. The defense industry workforce grew from 14,000 in 1966 to 34,000 in 1972 and continued growing after 1973 Yom Kippur War to bring total employment (including subcontractors) to 63,000 in 1987. By then, it comprised more than 4% of the country's total workforce and 20% of the industrial labor force.[27]

The policy of creating a defense industry based on homegrown innovation encompassed not only the array of government and private-sector contractors, but also the army itself, the universities and research institutes, all of which enhanced the foundations of academic-industrial cooperation that would become a hallmark of Israel's technology industry. The zenith of the ambitious defense program was marked by the Lavi fighter jet program, which was launched in 1980 with significant American financial aid and encompassed a wide range of technologies. But the Lavi was an exception to the rule: The small size of Israel's domestic defense market, a highly competitive and politicized export market in which Israel would have trouble competing against American and European companies, as well as the limited industrial capacity of the domestic defense industry, meant that the drive to become more militarily self-sufficient focused on electronics, communications, and computers rather than platforms such as tanks or planes.[28] Thus, many of the engineering skills and technology that went into military R&D could rapidly find applications in the civilian sector where they would later manifest themselves in the first wave of Israeli start-ups, which focused on communications technology and network security and eventually made itself felt in more distant applications, such as medical electronics.

The impact of the embargo and the drive for arms self-sufficiency cannot be underestimated. While Israel had many of the technological resources and the human capital to develop a high-tech industry, it lacked the free market environment that might have put this to use in civilian applications before the 1990s. Moreover, there was no global market for technology and communications structured in a way that small Israeli companies stood a chance of competing in an industry of big vertically integrated companies, government monopolies, and heavily regulated markets. This would change in the 1990s, but in the late

1960s and 1970s, it was the urgent requirements of national security rather than the market that served as the driving force for innovation and development. Ordinary capitalist/entrepreneurial incentives were not the driving force, but a sense of urgency and mission, both on a personal and on a societal level, borne of crisis (a factor discussed in Chapter 8). It is true that in the 1980s, defense exports, mainly of electronics and communications equipment, had become a major component of the economy, but the export aspect of the defense industry was a secondary outcome of a successful drive to develop new technology and the need to create the economies of scale needed to produce it at reasonable cost. Israeli companies were motivated by national security and then exploited the market opportunity that followed. Thus, although Israel's long history of wars and its perpetual war-footing would appear to have little relevance to the rise of its knowledge economy—or perhaps even act as a deterrent—military needs were in fact the primary driver for innovation until the 1990s.

Following the 1973 Yom Kippur War, Israel entered a period of slow economic growth and accelerating inflation that was reaching into the triple digits by the middle of the 1980s. There were multiple factors at play, none of which have direct bearing on the knowledge economy that would later emerge. However, broadly speaking they included a sharp rise in defense spending following the war that was later accompanied by increased social welfare spending, all of which conspired to create unsustainably large fiscal deficits. More fundamentally, the business sector dominated by the state and Histadrut could no longer meet the need for innovation and efficiency as the economy developed and grew more sophisticated. The era became the mirror opposite of the pre-state situation, where an undeveloped economy had attracted immigrants with educational and skills in excess of demand; now, Israel was producing a pool of human capital with insufficient domestic opportunities to apply it and benefit from it. Emigration had been a significant phenomenon all through Zionist history, but the 1970s and 1980s saw a rise in the rate. More significantly, large number of native-born were Israelis leaving, rather than recent immigrants with shallower roots in the country. By 1990, the number of Israelis living abroad had reached 6.2% of the population, a disproportionately large number of them highly educated and skilled people who found that they could not exercise their talents at home.[29]

All of these idologies and process over the previous century served as the deep foundation for the creation of a knowledge economy specific to Israel. However, the start-up economy only was able to emerge in the 1990s due to a consequence of domestic and global developments.

The first was the program of economic stabilization and reform undertaken starting in 1985. In its initial stage, it entailed reducing the fiscal deficit, export subsidies and import duties, suspending cost-of-living allowances, and instituting price controls, all with the immediate aim of bringing down inflation. A longer-term program in the years that followed undertook to privatize Israel's biggest companies (though to this day not the military industries), pare back government intervention in the economy, open up sectors such as telecommunications, finance, and transport to competition, and liberalize foreign trade. From a peak of 400% in 1984, inflation trended down to 20% in 1989 and 10% by 1996, and to the low single digits by the early 2000s. Given its relative isolation from the rest of the economy, the Israeli technology sector was affected less by the reforms of the 1980s and 1990s than other segments of the economy, but they undoubtedly improved investor confidence in Israel generally and created a more attractive environment for Israelis to remain to live and work in the country.

The next factor was the shutting of Lavi project in 1987 under pressure from Washington (as well as the Israel Defense Forces itself, which preferred to buy from the USA than to finance domestic R&D). While the cancelation of the Lavi was a big blow to the domestic defense industry, the industry had already begun shrinking in the decade before as the USA emerged as both a reliable and a generous ally, providing an annual $3 billion or more of financial assistance as well as access to some of its most advanced weapons technology. The effect of the cutbacks was to trim the workforce in the state-owned defense industry alone from 43,700 in 1985 to about 23,000 in 1997.[30] The Lavi's cancellation and the general contraction of the defense industry drove many of those engineers into the civilian sector at a particularly propitious moment, which constitutes the third major factor in the rise of Israeli high technology.

Starting in the USA and later spreading to Europe and much of the rest of the world, the decades-old monopoly on telecommunications services was gradually broken over the course of the 1980s and 1990s as technological innovations undermined the argument that telephone service was a so-called natural monopoly. In the USA, the process began with a 1984 consent decree requiring AT&T to divest its regional operating companies and allow competition in long-distance calling, spurring competition from firms such as MCI Communications and Sprint Communications. Competition and the entry of new firms into the market was re-enforced by the 1996 Telecommunications Act, which allowed long-distance companies, as well as cable television providers and others, to enter the local telephone business. Similar upheavals occurred in Europe, starting in 1985 with the European Community's Liberalization Directives and re-enforced by the 1996 Full Competition Directive in 1996, which required all member states to have a completely liberalized telecommunications market within two years. If Asia lagged behind the USA and Europe in deregulation, it made up for that by rapid economic growth and industrialization, which in turn increased demand both in terms of spending and sophistication for telecommunications services.

Deregulation was made possible by technology, mainly digitalization, but new technologies themselves also served as drivers in their own right. Personal computers for home and business use became widespread over the course of the 1980s, and in the following decade, cellular telephony and the Internet became mass-market phenomena. This not only created vast new markets but fragmented industries of multiple competitors and products comprised of outsourced components. In information technology, a handful of mainframe manufacturers who largely developed and made their own hardware and software quickly gave way to the personal computer makers who relied on open standards and a network of suppliers and created products compatible with third-party software and peripherals.[31] In telecommunications, the industry evolved from a small number of large makers of fixed-line telephony equipment selling to a limited number of service providers to an eclectic and dynamic market of mobile telecommunications, cordless telephony, trunking, and paging services. The convergence of telecommunications and information technology as the decade progressed created even more opportunities.

For the first generation of Israeli start-ups that were created in the 1990s, the market opportunities were seemingly boundless. The combined effect of deregulation and the rise of the Internet were to expand a telecoms universe once populated by a limited number of service providers (many of them state-controlled and rigidly managed) and equipment makers. The changes over the final two decades over the twentieth century created new product and service categories, markets and players and, indeed, served to enlarge the share of communications and information technology both in terms of time and spending for consumer and for business. As the supply of human capital and technology was becoming available in Israel, demand globally was exploding. Israel's industrial exports rose from $7.7 billion at start of the 1990s to $20.9 billion at the end of the decade as the share of high and medium-high technology exports grew to account for 86% of the total in 1999 from 51%.[32]

ECI Telecom, one of the first-generation technology companies, in many respects encapsulates the transition Israeli companies made amid the twin upheavals in the domestic defense sector and the global telecommunications industry. It had been formed in 1961 as a defense contractor but by the late 1970s, it began applying the technology it had developed to the civilian sector, most notably a telephone line doubler that allowed two-way conversations over a single cable. There was virtually no domestic market for ECI's products, and selling equipment abroad was extraordinarily difficult for a small, unknown company based in a country not yet associated with technology and viewed as a security risk.[33] Nevertheless, the company did win contracts from Deutsche Telecom and others, due to its proprietary technology and by concentrating on niche markets that were ignored by larger competitors, such as France's Alcatel. If ECI is unusual in the Israel rubric, it is because it remained largely focused on big telecoms operators even as the telecoms market began to diversify. But more importantly the company anticipated and successfully made the transition to digital technology

The fourth major factor—and one that is frequently overlooked in the growth of Israel's high-tech economy—is the gradual development of peace between Israel and its neighbors. Starting with Egyptian President Anwar Sadat's historic visit to Jerusalem in 1977, the process has been painstaking and slow, marked by severe setbacks and remains far from complete to this day. But the process has lifted the existential security threat that hung over Israel during its first decades and signaled an admission by the Arab world, however reluctantly, of the country's existence. Israel has been at peace with Egypt since 1979 and with Jordan since 1994. Critically, no final agreement has been achieved with the Palestinians and there have been surges of violence, during the Second Intifada (2000–2005) and Israel's periodic wars with the Palestinian Hamas organization (2008, 2012 and 2014). But the 1993 Oslo Accords were an important step in winding down the war between the two sides by establishing the outlines of an eventual settlement, allowing the Palestinians to create the rudiments of a state in the West Bank and Gaza, and creating a channel for the two sides negotiate.[34] Israel remains in a state of war with its two other neighbors—Syria and Lebanon—but neither country as of 2016 presents a strategic threat. Israel's enemies in terms of military threat are no longer the Arab states but non-state actors like Hamas, Hezbollah, and Hamas, who have forced Israel into frequent hostilities. But these are short wars in which the economic impact is small and the outcome of a stalemate or an Israeli victory is clear. The sole existential threat facing Israel as of 2016 is from Iran, which is 1800 kilometers away from Israel. Its modus operandi in terms of the cold war conflict it has been engaging in with Israel is via proxies like Hezbollah and Hamas and the threat of missile attack. While both are strategic threats, neither of which has fundamentally affected the way the Israel or world business community perceives political risk in Israel. In sum, Israel is not a member of the Middle East community of nations and its relations with its neighbors are far from normalized, but the fact of its existence is respected and at a time of increasing regional disarray it has even become a welcomed source of stability. The conflict between Israel and its Arab neighbors is no longer the focus of Israel or the Arab world's attention.[35]

The direct economic impact of the peace process for Israel has been marginal, and for the high-technology sector even more so. The economies of the confrontation states are small and underdeveloped markets. Even in those countries where bilateral trade relations with Israel exist, popular attitudes toward Israel remain hostile, discouraging business dealings and tourism. For Israel's technology companies, which are in the main acting as suppliers of software, components, and sophisticated equipment to other companies, there are virtually no potential customers in the region. However, the indirect impact of the peace process has been quite significant. It has enabled Israel to substantially reduce its defense spending over the past three decades as the strategic threat has gradually declined. In the 1973–1977 period, security expenditures accounted for in excess of 25% of gross domestic product, a result of the re-arming process following the 1973 Yom Kippur War, which had cost the army considerable losses in materiel. By the end of 1980s, the rate had declined to about 15%. Over the 1990s it took another step down to just over 10% of GDP in 1990–1996 and to a 7–9% range in the second half the decade, where it has remained since.[36] The decline in relative spending on defense enabled Israel to spend more on infrastructure, social welfare, health, and education and to reduce what had once been an immense tax burden. In 2015, for example, the tax component of Israeli labor costs (the tax wedge) was just 21.6%, compared with 29.6%

15 years earlier and below the 2015 average of 35.9% on average for countries belonging to the Organization for Economic Cooperation and Development.[37]

The peace process, together with the end of the Cold War, also had the effect of helping to undermine the Arab boycott against Israel, starting with the abortive 1991 Madrid peace conference and continuing with the Oslo process two years later. While a ban on doing business directly with Israel remains in force in much of the Arab world, it is far less important than the ban against third-country companies that do business with Israel, which has largely disappeared.[38] For Israeli companies, the weakening of the boycott created new customers and suppliers in the West and in the emerging Asian economies. This process was helped by Israel's growing acceptance in the world community and the establishment (or re-establishment) of diplomatic and trade relations with countries that had sided with the Arabs during their decades-long confrontation with Israel. Among the most important, China recognized Israel in 1992 and India formalized relations the same year, while Russia restored relations severed after the Six-Day War in 1991. Multifaceted ties with Turkey, involving political, economic, and military cooperation, deepened during the decade.

Finally, the security uncertainty that had suffered Israel from the time of its founding has gradually dissipated as its place in the region has become more secure. Its victory in the 1967 Six-Day War marked the first time that the Jewish state was regarded by both its friends and its foes as militarily undefeatable. Nevertheless, it remained shadowed by frequent wars, chronic terrorism, and political attacks on its right to exist and/or its acceptance in the community of nations. As the weakening of the Arab boycott and the establishment of diplomatic relations during the 1990s demonstrated, the peace process enabled to Israel to create a network of diplomatic, business, cultural, and other ties with the wider world that has re-enhanced its legitimacy.

Finally, Israel's human capital resources were substantially expanded by the wave of aliyah from the former Soviet Union. From the end of 1989 until the end of 1997, some 711,000 people arrived, expanding the size of the labor force by 16%. But more than that, the immigrants who arrived were unusually accomplished in terms of schooling and professions. In the very first wave, through the end of 1993, they had an average of 14.5 years of schooling and half held academic or managerial positions before arriving in Israel. Of those, 57,400 were engineers (compared with 30,200 resident in Israel in 1989) and 12,200 were medical doctors (compared with 15,600 resident).[39] The proportion of the Jewish population with 16 years or more of schooling grew from 12.2% in 1990 to 17.5% in 1999.[40] In point of fact, many of the most skilled immigrants had training and education in disciplines irrelevant to the needs of Israel's emergent high-tech sector, but the industry organized a massive retaining program, involving some 20,000 people, to turn civil and agricultural engineers into hardware and software specialists.[41]

A second immigration that occurred during the decade of the 1990s was smaller, but in many ways just as significant as the wave of Russians. Many of the Israelis who left their country during the economy's "lost years" of the 1970s and 1980s found themselves in Silicon Valley, joining emerging culture of start-up companies and venture capital, acquiring market intelligence, and developing personal and professional connections with American business people. This connection was critical as it added to the local pool of engineering and scientific talent a cadre of entrepreneurs who had a deep understanding of the US market, which was the undisputed world leader in determining the direction of emergent information and communications technology. In the 1990s, Silicon Valley, not only served the needs of a rapidly changing market but anticipated them, and often drove them, with technology innovations that created their own new product and service segments. Without an influx of Israelis who were at home both in their country's emerging start-up culture and the ways of Silicon Valley, Israeli start-ups, located thousands of miles away from the center of global technology innovation and from their US customers, would have had enormous difficulty keeping up with the pace of the business and technology developments taking place. Thus, as the Israeli economy stabilized and recovered in the 1990s and the high-technology industry began to develop, this diaspora of Israelis would become a critical asset. Many of those diaspora Israelis returned home to form new companies, established local research and development centers for their American employers, and/or invested in companies in Israel, providing strategic guidance on the board level. As a result, Israeli start-ups of the 1990s enjoyed critical links with the Silicon Valley from the outset, creating businesses that straddled the two countries by combining Israeli engineering abilities with access to and knowledge of the American market. Typically, Israeli start-ups of the 1990s based their research and development operations in Israel and their sales and marketing in the USA as well as their legal domicile.

Notes

1. Yair Aharoni, *The Israeli Economy: Dreams and Realities* (New York, 1991), 58.
2. The history of Jewish immigration to Palestine is customarily divided into waves of immigration (*aliyot*), each with its own social and economic characteristics. The first lasted from 1882 to 1903.
3. Ibid., 56.
4. Walter Laquer, *History of Zionism* (New York, 1972), 285.
5. Oz Almog, *The Sabra: The Creation of the New Jew* (Berkeley, CA, 2000) describes the history and characteristics of this phenomenon, which emerged as the dominant ideal of culture in the years 1930–1960. Its influence is described in depth in Chapters 3 and 8.
6. David Vital, *Zionism: The Formative Years* (Oxford, 1982), 390.
7. Laquer, *History of Zionism*, 310–311.
8. Yakir Plessner, *The Political Economy of Israel* (Albany, NY, 1994), 72.
9. Aharoni, *The Israeli Economy*, 67.
10. S. Ilan Troen, "Higher Education in Israel: A Historical Perspective," *Higher Education*, Vol. 23, No. 1 (January 1992), 47.
11. Ibid., 46.
12. Ibid., 52.
13. Ibid., 56.
14. Ibid.
15. Amir Ben Porat, *State and Capitalism in Israel* (Westport, CT, 1993), 36.
16. Aharoni, *The Israeli Economy*, 73.
17. Ibid., 169.
18. Troen, "Higher Education in Israel," 59.
19. Israel Council of Higher Education, Students in Institutions of Higher Education (Table 1).
20. Uri Cohen, "The Hebrew University and Ethno-Class Stratification in the First Decade," in Avi Bareli, Daniel Gutwein and Tuvia Friling, eds., *Society and Economy in Israel: A Historical and Contemporary Perspective* (Hebrew) (Jerusalem and Beersheva, 2005), 233.
    Cohen sees this change as an inter-ethnic struggle designed to ensure the dominance of Israel's European-origin population, its culture values.
21. S. Tevet, "What Happened to the Palmach Youth?'" *Ha'aretz*, May 14, 1965, quoted in Ibid., 241.
22. Ibid., 251.
23. Dan Brenitz, *Innovation and State: Political Choice and Strategies for Growth in Israel, Taiwan and Ireland* (New Haven, CT, 2007), 42.
24. Ibid., 42.
25. Ibid., 52–53.

26. Sharon Sadeh, "Israel's Beleaguered Defense Industry," *Middle East Review of International Affairs*, Vol. 5, No. 1 (March 2001).
27. Ibid.
28. Ibid.
29. Yinon Cohen, "Israeli-born Emigrants: Size, Destinations and Selectivity," *International Journal of Comparative Sociology*, Vol. 52, No. 1–2 (2011), 45–62, 47.
30. Sadeh, "Israel's Beleaguered Defense Industry."
31. Henry S. Rowen, "Serendipity of Strategy: How Technology and Markets Came to Favor Silicon Valley," in Chong-Moon Lee et al., eds., *The Silicon Valley Edge: A Habitat for Innovation and Entrepreneurship* (Stanford, CA, 2000), 194–198.
32. Yaacov Kop, "The Israeli Economy in the Fifth Decade Following D. Patinkin's The Israeli Economy in the First Decade," in Avi Bareli, Daniel Gutwein, and Tuvia Friling, eds., *Society and Economy in Israel: A Historical and Contemporary Perspective* (Hebrew) (Jerusalem and Beersheva: Yad Yitzhak Ben-Zvi and Ben-Gurion University Press, 2005), 58.
33. David Rosenberg, *Cloning Silicon Valley* (London: Pearson Education, 2001), 103–104.
34. Zaki Shalom, "The Peace Process: From Oslo Parameters to Unilateral Actions," *INSS Insight*, No. 267, July 12, 2011.
35. Amos Yadin, "Five Years Back and Five Years Forward: Israel's Strategic Environment in 2011–2015 and Policy Recommendations for 2016–2020," in Shlomo Brom and Anat Kurz, eds., *Strategic Survey for Israel 2015–2016*, 165–166.
36. Central Bureau of Statistics, General Government Defense Consumption.
37. Organization for Economic Cooperation and Development, *Taxing Wages 2014–2015 and Recent Trends* (Paris, 2016), 18.
38. Martin A. Weiss, *Arab League Boycott of Israel* (Washington, December 19, 2013), 2–3.
39. Eckstein and Weiss, "The Integration of Immigrants from the Former Soviet Union in the Israeli *Labor Market*," 2.
40. Yaacov Kop, "The Israeli Economy in the Fifth Decade," 36.
41. *Ha'aretz*, May 30, 2012.

CHAPTER 3

Israel as a Knowledge Economy I

Much as France is associated with luxury goods, Italy with fine fabrics, the Persian Gulf with oil, and (in times past) Detroit with automotive industry, Israel has become linked with high technology. If it doesn't rival California's Silicon Valley in size and scale, Israel has unarguably emerged as one of the few places outside the USA where a critical mass of engineers and entrepreneurs, companies, financial institutions, and universities generate a constant stream of innovation in the core fields of global technology, namely computers and communications. But unlike most industrial clusters, high-tech centers—or Silicon Valleys—involve more than a concentration of human expertise and physical infrastructure. As an industry, high technology is characterized by an organizational structure, values, attitudes, and processes that distinguish it from other sectors of the industrial economy. The technology business is characterized by flat organization, a culture of knowledge sharing between organizations and inside the organization, high levels of business risk, and the ability to exploit and adapt to rapidly changing markets and business models. If other industries over the last two decades have adopted some of its mores, global high technology remains the exemplar. In that respect, Israel has not only developed a critical mass of technology companies but has adopted the Silicon Valley ethos in its most extreme form by creating an industry of pure technology and innovation virtually untouched by the structures and norms of older industries.

D. Rosenberg, *Israel's Technology Economy*, Middle East in Focus,
https://doi.org/10.1007/978-3-319-76654-6_3

Israel's high-tech industry does not engage in manufacturing like Taiwan, Korea, and many other Asian economies. Nor has it created large multifaceted companies like Apple or Samsung that engage in the full array of business operations from research and development to marketing and engineering to finance and manufacturing (or overseeing a network of outsourced manufacturing). Nor does it have companies that dominate mass consumer markets and employ vast logistical operations, like Google, Amazon, Facebook or Alibaba. All that that would require a set of managerial and logistical skills and cost advantages the country lacks. Israeli companies have created a handful of significant, market-pioneering products, such as Internet chat, flash memory devices, voice over Internet protocol (VOIP), and certain computer firewall technology. Israeli-based engineers at Intel have been responsible for developing the 8088 computer processor, MMX technology, and the Centrino mobile technology. But Israeli companies have been on the whole been unsuccessful in capitalizing on their innovative capacities by building large, sustainable businesses from them. Rather, Israel's reputation lies in its ability to generate original new technology by way of start-up companies, most of it for the telecommunications and information technology applications and for use by business or inside the products of other companies. Over the last two decades, its investors, entrepreneurs, and engineers have adapted an idea that originated in the USA in the 1960s and 1970s—forming companies based on the intellectual capital of its founders and employees and backed by private investment with the aim of developing an original new service or product—and employed to an extent unrivaled anywhere in the world outside of Silicon Valley itself.

A few figures on the economy illustrate the phenomenon. In 2013, Israel spent 4.21% of its gross domestic product on research and development, the largest of any economy in the world and more than double the 1.91% for the 28 countries of the European Union.[1] In the years 2009 through 2014, Israel generated between 650 and 1005 start-up companies annually.[2] On a per capita basis, its technology companies raise more venture capital than the leading economies of Europe and the USA.[3] In terms of venture capital as a percentage of gross domestic product, a key measure of high-tech prowess, Israel towers over the rest of the world at 0.38% in 2014 and close to 70% of it directed at early-stage start-up companies.[4] In the USA, which No. 2 in the world by that measure, VC was just 0.28% of GDP and only a third of that went into early-stage firms. In 2010, Israel had the second-highest number of patent applications with the US Patent and Trademark Office on a per capita basis after the USA itself; in the European Union, where Israelis are less likely to seek intellectual property protection, Israel still ranked No. 5.[5] As an exporter of information and communications technology, Israel had a 2.7% share of global exports in 2011, ranking it No. 10 in absolute terms (behind France, ahead of Sweden), thanks to the presence of so many multinational research and development centers in Israel.[6]

Israel's achievements in creating a knowledge economy are remarkable given that during the 25 years that saw the rise of its technology industry the country contended with considerable military, political, social, and economic pressures. As it turns out, however, far from deterring the growth of the knowledge economy, they arguably contributed to it, certainly in regards to the form it has taken as a center for innovation.

In the military sphere, the era began with the First Gulf War (1990–1991), a conflict in which Israel was not a participant but nevertheless became the target of Iraqi missile attacks. The extent of the material damage was relatively small—42 Scud missiles reached Israel, killing one person and injuring 230 over seven weeks—but the attacks signaled a new era in the wider Israeli-Arab conflict. With the exception of the 1948 War of Independence, Israel's wars had been fought by conventional armies far away from its cities, industries, and farms (at least by Israeli standards). But, starting with the Gulf War, that was no longer the case. The 2006 Second Lebanon War, which pitted Israel against the Lebanese Shiite militia movement Hezbollah, saw the country's north effectively shut down for the month of fighting, while Hezbollah launched some 4000 missiles aimed mostly at civilian targets. Since 2009, Israel has fought three short wars with Hamas each lasting from one to seven weeks characterized by sustained rocket attacks that have reached increasingly larger portions of the country as Hamas improves the range and accuracy of its arsenal. In between Israel has had to cope with frequent small-scale rocket attacks.

However, neither the direct damage of the missile wars nor the political, psychological, and economic risk they posed to Israel has had the kind of long-term impact on economic performance one would have expected. The Second Lebanon War established the pattern: It caused a decline in output during the fighting that was rapidly offset in succeeding quarters.[7] Looking at it from the perspective of physical damage and cost of lives, the fleeting economic impact of these wars understandable: More than a decade of Hamas rocket attacks—mostly done with home-made projectiles with very limited range, payload, or accuracy—have killed only 33 Israeli civilians as of mid-2014.[8] The nature of the rocket attacks is as much about creating fear and uncertainty than about causing death and destruction, but even by that lesser standard their impact on Israel's economic life (business confidence, entrepreneurial activity, and investment) over the long term has been minimal.

This is not to say that in the absence of war, all of these parameters would not have been better. At least one study suggests that for every additional percentage point Israel's defense burden rises, the economy's growth rate is trimmed by 0.33 percentage point—a substantial cost given an average annual GDP growth rate of 4%.[9] As much as the military's demand for technology has enormous spin-off effects on domestic industry, it deprives the private sector of financial and human resources to an even greater degree. However, this weight has almost certainly eased in recent years as Israel's defense burden has fallen, reducing its overall negative impact on the economy, while the army's growing reliance on technology has enlarged the spin-off benefits to the country's high-tech sector. In all events, the impact of war and terrorism (as distinct from military spending) over the last two decades has been surprisingly small and short-lived. Decades of living in a state of perpetual security tensions or frequent conflicts have inured Israelis to security risk, an issue discussed further in Chapter 8.

The second kind of security pressure Israel has experienced over the last 25 years has come in the form of widespread grassroots violence on the part of the Palestinian population in the West Bank. In contrast to the rocket wars, the Second Intifada (2000–2005) cost about 1100 Israeli (as well as 5550 Palestinian) lives and contributed to pushing Israel into its deepest-ever recession during the first years of the violence. Palestinian shooting and bombing attacks reached Israel's biggest cities and continued for years, undermining consumer, business, and investor confidence, and deterring tourism.[10] But the Intifada caused very little direct material damage and other factors contributed at least as much to the economic downturn, including the bursting of the late 1990s global technology bubble and recession in the USA. In all events, Israel's economy showed unusual resilience, resuming growth in 2003 and expanding at a rate in excess of 4% annually even as the Intifada continued to rage. While the experience of grassroots violence was certainly traumatic, it had run its course long before the Second Intifada died down, further evidence of Israeli resiliency in the face of security uncertainty.

Thus, while the country's perennial conflict with its Arab neighbors is regarded by most of the world as the foremost factor in Israeli life, it has gradually receded in the Israeli consciousness. The threat of a sustained missile war that causes significant human and material damage remains, as does the threat of another outbreak of grassroots violence in the West Bank. However, as much as they may lie deep inside the Israeli consciousness and affect the Israeli worldview, these threats do not affect day-to-day economic life (where they do play a role is in attitudes toward business and innovation, a factor discussed in Chapter 7. The lower threat level is evidenced by military spending and compulsory army service, both of which have fallen over the last two decades. Defense spending as a percentage of GDP was 5.6% in 2013, down from an average of more than 10% in the early 1990s and a high of 30.3% in 1975.[11] Rates of enlistment for young people remain high—and among women they have grown as more opportunities open up for them in military service. The overall decline that has occurred from 77% of young people in 2004 to 72% in 2016 is mostly a function of the growing ultra-Orthodox population, whose young men and women are nearly all exempt from service.[12] There is considerable public discussion in Israel about declining motivation among the young to serve in the army, but as the figures suggest there doesn't seem much evidence of that. Rather, the motivation has changed. Enlistees are less likely to see their service as protecting their country from imminent threat. Rather, it has become a rite of passage and for the best and most ambitious recruits a way to enhance career prospects by way of the prestige, training, and social networks that come with being in top technology, intelligence, and combat units.[13] In any case, the impact of army service on a person's life in terms of time has shrunk: Mandatory service has been reduced mostly recently in 2015 to 32 months from 36 for men. In 2015, only 26% of Israelis eligible for the reserve duty had done at least 20 days of service in the previous three years.[14]

In the realm of politics, Israel faces an unusual set of pressures related to its Israel's international standing and in extreme cases even its right to exist. Since it was created in 1948, Israel has struggled to win broad global acceptance as a member of the international community, a critical issue for a small economy reliant on international trade and investment and on foreign military and diplomatic support to cope with its fragile security situation. The level of acceptance has waxed and waned over the decades in the face of the Arab world's rejection of the principle of a Jewish state and the unresolved Palestinian issue. However, the last 25 years have on the whole been favorable to Israel. The end of the Cold War together with the advent of the Oslo peace process in the 1990s did much to end Israel's relative isolation, enabling it to restore relations with the former Soviet Bloc countries and most critically with the emerging powers of China and India. It also brought an end to the unofficial boycott of Israel by Western and Japanese companies, which had been anxious not to risk their much more extensive interests in the Arab world by doing business with Israel.

The result has been a rapid growth in Israel's trade and investment ties with the global economy over the last decade and a half. The phenomenon has encompassed Israel's both traditional partners, Europe and the USA, and more recently with the rising Asian economic powers, principally China and India. In the 1999–2013 period, Israeli exports grew 92.2%, much faster than the 79.6% average for all developed economies.[15] With the European Union, Israel signed a free-trade agreement in 1975 and broadened it to an association agreement in 1995; in addition, Israel was the first non-European country to join both the EU's Framework Program for Research and Technical Development and the European Organization for Nuclear Research (CERN) in recognition of the high level of its scientific and technology research. Israeli exports to EU countries rose 98% in 1999–2013, exceeding overall export growth, although the EU's share has declined in the face of more rapid growth to other markets. Since 1985, Israel has had an FTA agreement with the USA as well. Although trade growth in North America over the years 1999–2013 rose a more modest 61%,[16] bilateral economic ties run arguably more deeply than with Europe. US companies, led by Intel, are the largest foreign investors in the Israeli economy and Israeli start-ups usually choose America as their principal market, setting up offices and often moving their headquarters there in the process.

Asia has emerged in recent years as the third great pillar of bilateral trade and investment for Israel. Israeli exports grew 188% over the 1999–2013 period, and since 2012, foreign direct investment mainly from China and India has followed. Growing economic ties are, of course, a function of the rising global economic clout of Asian economies, but it is also due to Israel's innovative capacity. China, which began as a manufacturer of simple products and/or a subcontractor to Western companies, is gradually moving higher up the economic value chain, evolving from a focus on manufacturing to developing industries capable of producing cutting-edge products and services. One way to fast track that development is to in effect outsource R&D by investing, acquiring, and partnering with Israeli start-up companies. Israel in that respect is in a unique position—at once a promising source for the technology and, unlike the USA and Europe, neither an industrial power nor an aspiring one jealous of its intellectual property.[17] On a macroeconomic level, China's government needs technology that will help ensure food, clean water, and energy as its population grows more affluent, all areas where Israel has developed solutions. Likewise India[18] and Japan,[19] even though they are at very different stages industrial development, are looking to tap Israel's knowledge economy for much the same reason. Thus merchandise trade is less of a barometer of Israel's growing ties with Asia than in the developing network of investment and knowledge sharing. These have included extensive Chinese, and more recently Indian and Japanese, investment in Israel start-up companies,[20] tie-ups between Israeli and Chinese universities, as well as joint research and development programs between governments.

The one major exception to Israel's global opening has been the Middle East, where Israel has remained an outsider economically and politically even vis-a-vis Egypt and Jordan, the two Arab countries with which it has diplomatic relations. Decades of formal relations have done little to moderate hostile public opinion in Egypt and Jordan toward Israel, which has thwarted the development of normal trade, tourism, and investment relations. However, for the knowledge segment of Israel's economy, regional isolation has not been a significant factor given that the Arab world has no significant role in global information and communications technology industry. In any case, there is considerable anecdotal evidence that Israeli technology leaks into the economies of the Arab world either through unpublicized direct dealings or through the global supply chain.[21] However, for the rest of Israeli industry, regional isolation has imposed an opportunity cost by preventing companies from fully exploiting lower labor costs in adjacent Arab countries, depriving Israeli companies of nearby markets and denying Israel's economy access to the region's abundant energy reserves.[22]

Over the last decade or more, Israel's global opening has been marred by the stalled Palestinian peace process and the increased international scrutiny of Israel's military operations and human rights policies vis-a-vis the Palestinians. Both of have given impetus to a grassroots movement to impose boycotts on Israel that got under way in 2004. Yet, while the boycott, sanctions, and divestment (BDS) movement has elicited at times an overwrought response from the Israeli government, the fact is BDS has failed to impose economic costs on Israel, in spite of extensive media coverage and a few isolated incidents of pension funds and companies undertaking very limited divestments or pulling out of contracts.[23] In Europe, where the BDS movement is strongest and governments have taken the firmest stand against doing business with Israel's West Bank settlements, Israeli exports doubled their annual average from $7.8 billion in 1995–2004 to $15.6 billion in 2005–2013, despite the sharp downturn in trade during the 2008–2009 global financial crisis.[24] In the USA, the BDS movement has made far less headway in the public consciousness and Washington has done nothing to deny Israel trade benefits to the settlements.

In all events, the BDS movement faces a serious obstacle in mounting a consumer-based boycott because few Israeli companies sell consumer products abroad. The movement has thus been forced to focus its attention on the few small and medium-sized companies that are vulnerable, such as SodaStream International (a maker of home carbonation systems) and Ahava (which makes skincare treatments). There, BDS has had some limited successes, but not on the scale that has imposed costs on the Israeli economy. On the other hand, Israel's knowledge sector is largely immune to consumer boycotts because its companies produce mainly intermediate products that are sold to other businesses and sit (usually anonymously) deep inside end-user products or computer networks.[25] If there is a risk for Israeli technology, it is because a growing number of companies are entering the consumer sector (as discussed in Chapter 6), but as of 2017 that hasn't happened on a scale to elicit the attention of boycotters. Where Israel is at risk to boycotts and sanctions is in the event a government or supra-government opts to take action, as the EU has done on a very limited scale. Using 2012 economic data, the Israel Finance Ministry estimated the impact of a "voluntary" boycott by retail chains in the Europe Union would amount to just 1.1 billion shekels or 0.5% of gross domestic product.[26] By comparison, a loss of 20% of Israeli exports to the EU, a scenario realistically only possible if a boycott is imposed by the EU or member governments, the lost exports would amount to 19.8 billion shekels or 10.1% of GDP.

For Israel, Asia offers a way to hedge the risk that boycott pressure may one day narrow business opportunities in Europe, if not the USA.

Although such views may prove in the long run to be naïve, in Israel the perception is that in Asia there is little grassroots public interest in the Israel-Palestinian dispute and that Asian policymakers do not mix human rights considerations with strategic economic and political goals. As well, Israeli policymakers correctly believe that the country can leverage its technology assets both economically and politically in ways it could never do through ordinary two-way trade because Israel by that measure is too small to factor into the policy strategies of the big Asian powers.[27]

The third pressure Israel has contended with over the last 25 years came with the vast wave of immigration from the former Soviet Union, which began in 1989 and peaked in the early part of the 1990s. Israel's population grew by one-fifth over the course of the decade, about half of that increase due to immigration,[28] creating a huge burden on the economy as it struggled to rapidly ensure housing, infrastructure, and jobs. The unemployment rate climbed from 6.4% in 1988 to a peak of 11.2% in 1992, but like the shocks administrated by upsurges in war and terrorism, the Israeli economy proved able to absorb the newcomers. Over the 1990s, GDP grew 60%, or an average rate of 4.8% annually (1.9% on a per capita basis).[29] Not only did the jobless level decline as the immigrants were absorbed in the workforce, their unusually strong educational profile provided a reservoir of skills from which Israel's emerging high-tech sector would draw, as is discussed in more detail in Chapter 7.

It was by no means assured that this mass influx of Russians would be absorbed so successfully rather than becoming a source for social discontent amid heightened competition for jobs, services, and government aid. The migrants from the former Soviet Union entered into a delicate constellation of ethnic, religious, and class divisions in Israel, which constitutes the second part of the demographic challenge that has faced Israeli society over the past 25 years.[30] The smooth functioning of Israeli democracy and its adherence to Western norms of freedom and rule of law (at least inside its pre-1967 borders) has masked deep political and social divisions. The political divisions are expressed principally in fundamental disagreements over making peace with the Palestinians and retaining the West Bank and its Israeli settlements, which in turn reflect a deeper ideological division about Israel's place in the Middle East and the wider world—whether it can reach a true acceptance from the Arab world and whether it should aspire to be a welcomed member of the international community or accept its "outsider" status as a historical destiny. These controversies are so critical to the existence of the state and speak so deeply to Israelis that they have had the effect of crowding out many of the other issues that preoccupy Western democratic societies in the 2000s, such as the status of women and gays, economic policy, and the environment. The differences have manifested themselves in the wrenching debates in parliament and on the street over Oslo peace process in the 1990s and the various attempts to revive it over the past decade as well as the dismantling of settlements, most notably the 2005 Gaza Strip evacuation and periodic efforts by the government to dismantle unauthorized outposts in the West Bank.

These ideological fissures are exacerbated by religious, ethnic and national divisions. Due to its history, these social divisions cut across an unusually large number of directions for a country of just eight million people. The deepest of these is between Jews and Israel's Arab minority, a division that encompasses language, religion, socioeconomic status, and national identity. Among Jews, there are multiple divides, the first being between those who trace their origins to Europe and North America (Ashkenazim) and those of North African and Middle East origin (Mizrahim). These began with cultural differences but became exacerbated by the lower socioeconomic status of Mizrahim and eventually came to be expressed in political differences, with Mizrahim favoring right-wing and religious parties over the center-left parties that dominated Israeli political life in the country's first three decades. Among Ashkenazim, Russian immigrants constitute a distinct bloc because their numbers are so large and because they arrived relatively recently and are only gradually assimilating into Israeli society. A second big division is based on religion among the Jewish population. Secular and traditional Jews form one bloc, national-religious Jews (or what Americans would call Modern Orthodox) a second, and the ultra-Orthodox an increasingly large and influential third bloc. These differences manifest themselves less over cultural and ideological issues such as abortion, evolution, and the like, as they do in America, but primarily more over issues of public religious observance and the role of the state in enforcing it, especially observance of the Jewish Sabbath.

If some of these fissures have eroded over the last two decades, with social statistics pointing to the gradual disappearance of cultural differences and income gaps in the case of Mizrahim and Russians in particular (see Chapter 9), in the public mind they remain very much alive and manifest themselves in political parties that speak in the name of the different groups—Shas for Mizrahim, Yisrael Beiteinu for Russian immigrants, Balad and the United Arab List-Ta'al for Israeli Arabs, and to a lesser extent the Labor and Meretz parties middle class for secular Jews of European origin. The two populations that have so far failed to join the Israeli melting pot are Israeli Arabs and Haredim, who between them account for close to 30% of Israel's population. Each presents a special case discussed in Chapters 10 and 11.

That fact that Israel has a plethora of political parties is due its system of proportional representation, which awards parties seats in parliament based on their nationwide vote rather than by geographical constituencies. Although the system has been criticized for creating unstable governments, proportional representation has served as a mechanism for channeling political discontent by ensuring a wide range of groups each has their place in parliament and often the cabinet itself. Political violence does well up in Israel, most notably the 1995 assassination of Prime Minister Yitzhak Rabin, which came amid the ideological conflict over peace with the Palestinians and the future of the West Bank. But considering the depth of political and social fissures in Israeli society and the pressures of being on a perpetual war footing, the instances of political violence are rare, and much of the credit must go to a political system that serves as out an outlet for grievances. The downside of the system is that it creates perpetual political instability that has made pursuing consistent long-term policies very difficult. No party in Israel's history has ever garnered enough votes to form a government on its own so that every government since the country's first elections in 1949 has been a coalition of multiple partners representing a wide range of interests formed via protracted and complex negotiation following elections. Unresolved divisions express themselves in frequent changes of government and policy, a phenomenon that has grown more pronounced over the last two decades. Among 19 parliamentary democracies surveyed by the Israel Democracy Institute, Israel ranked 13th for the average term for a prime minister since the state was established (3.8 years).[31] Moreover, the average term of a prime minister has fallen by more than a third in the 20 years after 1990 to 1015 days from 1581 before 1990. Between 1990 and 2015, Israel held parliamentary and/or prime ministerial elections nine times, with the party in power being returned to office in only a third of the votes.[32]

In spite of the instability at the top, the preoccupation of public debate with issues of national security, religion, and state and other purely political matters has served Israel's economic policymaking well by narrowing discussion to a small group of government officials, business leaders, and academics. Although this closed circle has at various times been penetrated by outsiders—most notably labor unions, the populist media, and briefly by mass rallies in the summer of 2011—this circle has been left relatively free to determine priorities and the direction of the economy. The thrust of that policy since 1985, and most notably after Benjamin Netanyahu became finance minister in 2003, has been to reduce the role of government in the economy through a multipronged strategy of fiscal discipline, lower taxes, reduced spending (particularly on social services and transfer payments), deregulation, and privatization of government-owned companies. Public-sector spending as a percentage of GDP has declined from an average of 50% in the 1996–2000 period to about 39% in 2015, the lowest level since the late 1960s.[33] Civilian spending dropped from 17.6% to 17.2% (with transfer payments pacing the decline, falling from 10.9% to 9.7%).[34] In spite of the Lebanon War and a subsequent rearming program, defense expenditures fell even more sharply—from 7.5% to 5.5%. Public debt to GDP, traditionally high compared to other developed economies, has been declining since 2003 when it was close to 100% to 62.1% in 2016[35] as the economy has grown and Israel was spared the vast stimulus programs and bank bailouts that many Western economies were forced to undertake in response to the 2009–2010 financial crisis. Since 2009, Israel has been a net creditor economy, with the surplus of assets to liabilities abroad steadily growing and reaching $68.5 billion at the end of 2015.[36] This makes Israel an outlier among OECD countries, against which it had historically had a much bigger public sector as measured by spending than the organization's average. However, public-sector spending has been on a steady decline since 2002 while in the OECD spending began to climb in the wake of the 2008 financial crisis and has remained at elevated levels since then. As of 2015, Israeli spending was about six percentage points less than the OECD average.[37]

Starting with the 1985 Economic Stabilization Program, the Israeli government has exited from or reduced its role in important segments of the economy, such as telecommunications, finance, pensions, agricultural marketing, and aviation, to name several of the most important. Likewise, the Histadrut labor federation, which acted in the first decades of Israel's history as a quasi-government body, shed its industrial, construction, and financial empire during the late 1980s and later as the main provider of health services through its Clalit health maintenance organization, which it spun off in 1993. Nevertheless, the government's role in the economy remains large because it retains control of such critical segments as ports and airports, electric power, railways, defense contracting, land, and water. These sectors operate as a monopoly, or near-monopoly, in the allocation and pricing of fundamental goods and services that have a multiplier effect across the rest of the economy. Apart from being monopolies, these state-owned enterprises effectively function as a government–Histadrut partnership—albeit a tense one—where the state maintains ostensive control but effectively leaves management to the labor unions. The role of powerful unions not only preserves these monopolies but prevents them from taking the kind of efficiency measures private enterprises routinely employ. Israel's public-sector workforce, which includes state-owned enterprises, has actually grown in the nearly three decades from 9.5% of the total population in 1985 to more than 15% in 2015.[38]

A case in point is the electricity sector, where Israel has lagged behind most OECD countries in opening up the market to competition in generation and transmission, despite reform proposals dating back to the early 1990s. State-owned Israel Electric Corp. (IEC) has gradually ceded some of its monopoly but at such a slow pace that by 2014, only 12% of total generating capacity was supplied by competitors while IEC continued to have exclusive control over transmission.[39] IEC itself is heavily in debt and has struggled to keep up with growing demand for power, such that reserve capacity was insufficient to meet the demands of a growing economy that is blocked for political reasons from connecting to other national power grids as a backup.[40] Consolidated Edison, which operates in the greater New York City area, generated approximately 62,000 million kilowatt hours of power or 25% more than IEC's 49,660 million, in addition to deliveries of natural gas and steam to customers. Yet Con Ed's labor force numbered just 16% more than IEC (about 14,600 versus 12,530).[41]

In short, Israel has employed the so-called Washington Consensus[42] over the last two decades at least in terms of fiscal policy, if so in terms of liberalization. The result was a period of virtually uninterrupted growth since the deep recession in 2001–2002. Over a 17-year period (1996–2013), the economy's average growth rate was 4%, compared with 2.15% for OECD members on average and 3.5% for the world.[43] However, the headline growth figures belie a less impressive performance after discounting for Israel's rapid population growth relative to other developed countries over the past 25 years. In terms of GDP per capita,

Israel has made some progress, although it remains well behind the 17 wealthiest economies belonging to the OECD.[44] In 1991, the gap between Israel and the top 17 was 33.2%. It narrowed the difference to 25.9% by 1996, but then lost ground again so that by 2003 the gap was slightly wider at 33.6%. In the decade since then, Israel has made steady but slow progress to narrow the difference to but as of 2013 it remained 22.4% even though Israel had enjoyed relatively strong economic growth while much of the OECD struggled through a deep recession and a sluggish revival in the aftermath.

One key reason for the lagging performance is Israel's low rate of labor productivity and its failure even to keep pace with growth in other developed economies. Measured in terms of hours worked, the gap between Israel and the upper half of OECD members grew 24% in 1991 to 37% by 2003. It subsequently narrowed, but in 2013 it was still 32.6%, wider than it had been 22 years earlier.[45] Ironically, one explanation for why productivity lags is the government's successful effort to bring more ultra-Orthodox men and Israeli-Arab women into the labor force. Both groups have traditionally suffered low rates of education and employment; as more and more of them find jobs, they are depressing average productivity levels. But that phenomenon has only emerged in recent years. Israel's lagging productivity has more fundamental causes. The first is that the country's talent is concentrated in the country's high-tech industry and in creating new technology for the global market instead of being employed in the service of the domestic market. That manifests itself in the fact that relative to their peers in the developed world, Israeli business, and even more so the Israeli government, does not make use of innovation, in terms of either deploying the most advanced machinery and equipment or putting into practice innovative management or productive processes. This issue discussed in more depth in Chapter 4, but it can be illustrated by one small example: The shortage of cybersecurity experts in business and government, which has exposed organizations to security risks, because so many are employed in high-tech companies offering cybersecurity solutions to the global market.[46] The second cause, one that has an especially profound impact on productivity, is the poor performance of Israeli schools, which produce graduates that on average have low levels of job-related skills and problem-solving abilities, a subject discussed in more depth in Chapter 7. Those with the highest skills gravitate toward the high-tech and to a few select companies in the export sector, the only areas where Israel as a result is globally competitive in terms of labor productivity. Thanks to its strong fiscal situation from years of adhering to the Washington Consensus, the Israeli government has the financial resources to upgrade its schools and has made some effort in this direction, but it has done nothing to fundamentally alter this dangerous situation for an economy so reliant on its knowledge resources.

To a large extent, this explains why Israel's top-line growth has failed to deliver commensurate benefits to the Israeli population in terms of reducing poverty, narrowing income inequality, and raising standards of living to the degree it should have. In the nine years to 2009, Israel's poverty rate grew sharply from 17.5% of the population to 25%, rising whether the economy was in recession or not, according to National Insurance Institute figures.[47] The rate has been in decline since 2009 but at 21.8% in 2013 it remained far above its level in 1998 despite 16 years of uninterrupted economic growth. In terms of income inequality, Israel had made some strides, with its Gini coefficient of income inequality showing an improvement of about 5.5% in 1999–2012 before taking into account income transfers and taxes.[48] A key reason for the decline in poverty was the long-term policy decision dating from the early 2000s to drive more Israelis into the labor market where the labor force participation rate had fallen to dangerously low levels. On the one side, the government reduced child allowances and other benefits that were seen as providing enough of a financial cushion to enable large segments of the working age population to remain out of the labor force. On the other side, Israel's tax regime was restructured to favor wage earners, especially the working poor, by raising consumption taxes, like the value-added tax, and lowering income tax rates, including the introduction of an earned-income tax credit. That had the effect of raising Israel's labor force participation rate even as the unemployment rate declined, with lower-income earners seeing the biggest employment gains after 2010.[49] Still, Israel remains among the most impoverished and unequal of OECD economies. Even after the improvement in recent years, Israel's poverty rate was nearly 60% higher than the 11.1% average for member countries in 2013 and was exceeded only by Mexico and Chile.[50] In terms of income inequality, Israel's was only exceeded among 34 OECD countries by the USA, Turkey, Mexico, and Chile.[51]

That said, it would be wrong to ascribe solely to government policy Israel's distressingly high poverty rate, which is influenced by social factors unique to Israel. One is the high level of immigration, although for the years surveyed by the NII that was not a major factor since the wave of Russian immigration had tailed off by the middle of the 1990s and the immigrants successfully absorbed into the labor market. The second factor is the voluntary poverty of it large and growing ultra-Orthodox sector, which has adopted a religious ideology over the last four decades of preferring religious study over paid employment, a phenomenon that will be examined in depth in Chapter 11. The third is the country's failure to integrate its Arab minority, which constitutes a fifth of the population, into the economy and the labor market, which will be discussed in Chapter 10. But another factor with direct bearing on its knowledge economy is that Israel's investment in its human capital is yielding poor results: Although the country spends a relatively large share of GDP on education, even after taking into account its relatively young population, scores in standardized international exams for math and science achievement consistently put Israel students close to the bottom of OECD countries,[52] which in turn is weighing on the economy's ability to generate productivity growth.[53] In fact, the growth area of Israel's labor market in recent years has been for jobs demanding low skills and education, which has been a boon for the country's low-income groups—Haredim and Arabs in particular. But the jobs pay relatively poorly, offer little career advancement, and provide poor employment security, all of which take a toll on productivity growth. Moreover, this stands in sharp contrast to the years 2004–2008, when the high-tech sector had led job growth.[54]

Although the Israeli knowledge economy has its own unique characteristics, in many respects it shares the same dilemmas that the world's other advanced economies do as they leave the old industry-centric model that predominated over most of the twentieth century. The assumption was that modern economies could grow indefinitely, delivering steady productivity improvements by deploying technology and a better-educated workforce. The result would be ever higher standards of living, shrinking rates of poverty, and narrowing income inequality. The rise of the knowledge economy has given serious pause to those assumptions in Israel and elsewhere. Israel's decisive move into the knowledge economy, exacerbated by its failure to fully exploit its knowledge assets as widely as it could or should have, has meant it has experienced the negative aspects of the knowledge economy faster and more severely than others.

Notes

1. OECD, *Science, Technology and Industry Scoreboard 2015* (Paris, October 23, 2013), 97.
2. Israel Venture Research, "Israeli Startup Success Report 1999–2014," January 28, 2015.
3. Ben Rooney, "How Entrepreneurial Is Europe?" *The Wall Street Journal*, November 11, 2013.
4. OECD, *Science, Technology and Industry Scoreboard 2015* (Paris, October 19, 2015), 176.
5. Daphne Getz et al., *Indices of Science, Technology and Innovation in Israel: An International Comparison* (Hebrew) (Haifa, 2013), 80–81. Israelis had 672 patents per million population at the USPTO, compared with 782 for Americans; in the EPO, the rate was 136, compared with 108 on average for EU members.
6. OECD, *Science, Technology and Industry Scoreboard 2013*, 217.
7. Bank of Israel, *Recent Economic Developments*, No. 116, July–December 2006, 1.
8. B'Tselem, "Attacks on Israel Civilians by Palestinians," July 24, 2014.
9. Mark Broude, Saadet Deger, and Somnath Sen, "Defense, Innovation and Development: The Case of Israel," *Journal of Innovation Economics and Management*, 2013/2, No. 12, 56.
10. Zvi Eckstein and Daniel Tsiddon, "Macroeconomic Consequences of Terror: Theory and the Case of Israel," December 10, 2003. The paper estimates that Second Intifada Terror reduced per capita output by 4% in the first three years.
11. Central Bureau of Statistics, Defense Expenditure in Israel, 1950–2013.
12. Nir Dvori, "Decline in Male Conscription Rates and Combat Motivations" (Hebrew), *Mako News*, November 16, 2016.
13. Reuven Gal, "Motivation Levels for IDF Enlistment over the Years," in *Military Service in Israel: Challenges and Ramifications, Memorandum* No. 159 (Hebrew), 56–57. Gal sees a rise in ideologically motivated army service, especially among Israel's Modern Orthodox Jewish population.
14. Judah Aryeh Gross, "Just a Quarter of All Eligible Reservists Serve in the IDF," *Times of Israel*, May 30, 2016.
15. Knesset Research and Information Center (Eyal Kauffman), *Study of the Possible Impact of an Economic Boycott of Israel* (Hebrew) (Jerusalem, December 31, 2014), 3.
16. Ibid., 4.
17. "Why Tech Transfer Is Key to Stronger China-Israel Ties," *Knowledge@Wharton*, August 13, 2015.

18. Tova Cohen, "Under Modi, Israel and India Forge Deeper Economic Ties," Reuters, November 19, 2014.
19. Ohad Cohen, "The Rise of Japanese-Israeli Economic Relations," *Times of Israel*, March 2, 2015.
20. David Rosenberg, "Chinese Investment in Israeli Tech Soaring," *Ha'aretz*, June 3, 2015.
21. Shuki Sadeh, "The Badly Kept Secret of Israel's Trade Throughout the Muslim World," *Ha'aretz*, January 19, 2012; Ronen Bergman, "How Israeli-Made Spyware Made It to the Arab World," Ynet, September 8, 2016.
22. Israel signed an agreement in 2005 to import Egyptian natural gas, which would have become by far the biggest single element of bilateral trade. Egypt abruptly canceled the gas accord in 2012 following the toppling of President Hosni Mubarak, citing a dispute over pricing. Egypt was desperately in need of gas for domestic needs, but it acted against a background of repeated attacks on the pipeline and broad public opposition to the agreement.
23. Knesset Research and Information Center (Eyal Kauffman), *Study of the Possible Impact of an Economic Boycott* (Hebrew) (Jerusalem, December 31, 2014), 11.
24. Ibid., 13. Among other things, the EU denies free-trade benefits given Israel to settlements and refuses to recognize Israeli veterinary supervision in the West Bank, effectively blocking farm exports from the territories. As of 2015, the EU has required products made in West Bank settlements to be labeled as such. These steps have had little impact because West Bank settlements account for less than 0.5% of Israeli merchandise exports.
25. Ibid., 11.
26. Shaul Amsterdamski, "Full Boycott Report Revealed: A Complete Boycott of Israel Would Cost 40 Billion Shekels a Year," *Calcalist*, July 7, 2015.
27. Ora Coren, "Israeli Economist: Business with China Should Hinge on Long-Term Ventures," *Ha'aretz*, December 27, 2013.
28. Yaacov Kop, "The Israeli Economy in the Fifth Decade Following D. Patinkin's The Israeli Economy in the First Decade," in Avi Bareli, Daniel Gutwein, and Tuvia Friling, eds., *Society and Economy in Israel: A Historical and Contemporary Perspective* (Hebrew) (Jerusalem and Beersheva, 2005), 34.
29. Ibid., 39.
30. Ibid., 35. Among the biggest demographic shifts, Israel's Ashkenazim (Jews of European origin) had lost their majority of Afro-Asian Jews in

the first four decades of the state, a status that was restored by the wave Russian immigration.

31. Shlomit Barnea, Shurik Dryshpitz, and Ofer Kenig, "Government Stability—Working Paper" (Hebrew), undated, 9–10.
32. Ariel Sharon, who had won the 2001 election for prime minister, reformed his government after the 2003 parliamentary elections that yielded his Likud Party a parliamentary majority. He subsequently left the Likud to firm the new Kadima Party, which went on the defeat the Likud in 2006 parliamentary elections.
33. Bank of Israel, *Annual Report 2015* (Jerusalem, April 3, 2016), 157.
34. Bank of Israel, *Annual Report 2013* (Jerusalem, April 2, 2014), 153.
35. Finance Ministry, "Accountant-General Publishes Preliminary Estimate of Public Debt to GDP in 2016 at 62.1%," January 22, 2017.
36. Bank of Israel, External Sector, International Investment Position, December 14, 2016.
37. Bank of Israel, *Annual Report 2011* (Jerusalem, March 28, 2012), 259, and *Annual Report 2015*, and 157.
38. Bank of Israel, *Annual Report 2015*, Total Number of Employees in the General Government and Business Sectors, 1969–2015 (Statistical Appendix). Israel's population was about 8.5 million in 2015 and its public-sector workforce numbered 1.31 million, compared with figures of 4.23 million and 402,900 in 1985.
39. Israel Electric Corp., *2014 Annual Report*, 24.
40. Bank of Israel, *Annual Report 2011*, 116.
41. Corporate financial reports, 2014, for Israel Electric and Consolidated Edison.
42. Although the term is subject to controversy, here the "Washington Consensus" refers to the broad definition, namely a policy strategy emphasizing free markets, fiscal discipline, liberalized trade, and privatization.
43. OECD, *Economic Survey for Israel, 2013* (Paris, 2013), 12.
44. OECD, *Economic Policy Reforms 2015: Going for Growth* (Paris, February 9, 2015), 223.
45. Ibid.
46. Amitai Ziv, "Israel at Risk Amid Shortage of Cybersecurity Experts," *Ha'aretz*, January 17, 2017.
47. National Insurance Institute, *Dimensions of Poverty and Social Gaps, 2013* (Jerusalem, December 2013), 58. Figures are for poverty after taking into account income transfers and taxes and include East Jerusalem.
48. National Insurance Institute, *Dimensions of Poverty and Social Gaps, 2012* (Jerusalem, December 2013), 39. Gini was higher after transfers and taxes, reflecting the shrinking role of social-welfare assistance.

49. National Insurance Institute, *Dimensions of Poverty, 2013*, 13–14.
50. Ibid., 42.
51. National Insurance Institute, *Dimensions of Poverty, 2012*, 35. Israel's rate for comparison purposes was 2012, versus 2011 for other OECD countries.
52. Ibid., 112. Israel's 2009 mean score on the Programme for International Student Assessment (PISA) exam was 459, compared with an average of 515 for the upper half of OECD countries ranked by GDP per capita and an average of 478 for lower-half countries.
53. Bank of Israel, *Recent Economic Developments*, No. 139 (Jerusalem, 2015), 6.
54. NII, *Dimensions of Poverty, 2013*, 13.

CHAPTER 4

Israel as a Knowledge Economy II

While Israel's so-called start-up nation has captured the popular imagination and made a significant contribution to the economy, the industry accounts for a relatively small part of the economy and has little connection with the rest of the country's business sector. The high technology industry employs relatively few people and those it does employ encompass a relatively small range of skills and professions. The information technology and communications sector, which includes local telecommunications companies, accounted for just 11% of GDP in 2014.[1] While high technology accounted for more than a third of all exports, a good measure of the industry's ability to develop and produce products and services for world markets, it only employed about 12% of the business-sector workforce.[2] The capital Israeli high tech employs mostly comes from overseas. Its products and services are almost wholly directed at foreign markets, principally America in its role as the global pacesetter for technology. Even the smallest and youngest start-ups find themselves establishing sales and marketing offices, and often headquarters, in the USA because the market is so critical not just to sales but to the very core of the enterprise. In more recent years, they have even come to rely more on overseas engineers as the supply and cost of domestic talent has become constricted. The disconnect from the surrounding economy more often than not reaches its logical conclusion for a great many Israeli start-ups by their being acquired by a foreign multinational and turned into local research and development units for their foreign owners. Because the Israeli technology sector sells few products or services to the local market it operates in an environment largely free

D. Rosenberg, *Israel's Technology Economy*, Middle East in Focus,
https://doi.org/10.1007/978-3-319-76654-6_4

of domestic regulatory or other governmental constraints or incentives; instead, its home is a global economy that is highly competitive and whose markets evolve at a breathtaking speed.

Israel's knowledge assets—its most highly skilled personnel, the deployment of technology hardware and software, and the use of innovative techniques of manufacturing, management and marketing—are concentrated in its start-up industry; the remainder of the country's business and government sectors—that is the great majority of the economy—make far less use of these. Thus, at home the business sector in contrast to the start-up sector is unable to deliver goods and services nearly as effectively to the market. Most Israeli business operates in a small and relatively closed economy, at least in terms of consumer goods and services, where economies of scale are unfavorable, competition is limited, and government oversight is often ineffective. One area where the problem is especially manifest is housing, which was one of the social justice protest's main agenda items. In the spring of 2011, home prices had climbed 8.8% after inflation from a year earlier and the ratio of home price to the average wage was at an historic high.[3] That trend threatened to saddle middle-class Israelis with a heavy burden of high mortgage payments for a decade or more into their future that would inevitably depress their other spending and perhaps their access many of the accouterments customarily regarded as a part of middle-class lifestyle.[4] The rise in residential real estate prices, which has continued in the years following the social justice protests, has been the result of a speculative bubble; rather, it was the consequence of fundamental distortions in the sector, where the state holds a near monopoly on land, excessive regulations and red tape slow the planning and approval process, and the construction industry is characterized by low productivity and high costs. The home-construction sector typifies the Israeli economy more than the start-up sector does. Thus, fully understanding the nature of Israel's knowledge economy requires examining the areas that employ the great majority of its labor force, comprise most of its output of goods and services, and play a far bigger role than the high-technology sector in forming Israel's economic and social parameters.

Israel's non-technology economy has been dominated by three phenomena over the last three decades or more. One is that the overwhelming presence of business groups, usually organized as pyramids of companies controlled by a single shareholder or shareholder group at the top. The second is restricted competition in most major markets for products and services. The third is the role of government, which has a baleful influence both as an inefficient regulator and administrator, and due its control of a handful of key monopoly industries. All of these problems are exacerbated by a poorly trained workforce, an issue discussed in Chapter 8.

The issue of business groups was little studied or the subject of serious policy discussions until the 2011 social justice protests, which brought to attention the social and economic dislocations that had emerged over the last two decades, even as the economy as a whole was growing. The business groups, personified by the individuals and families who run them (now etched in the popular mind as the tycoon class), came to symbolize the structural problems weighing on the economy. If, in fact, too much blame was laid on the tycoons, it is undeniable that they had grown too powerful, dominating markets for products and services and the allocation of capital. One measure of the extent of their control was their presence in the Tel Aviv Stock Exchange (TASE) where some 24 business groups controlled 136 publicly traded companies, or 23% of the total, as of September 2010.[5] In fact, that figure understates the presence of business groups: Measured by market capitalization, they accounted for 68% of TASE market capitalization, not including Teva Pharmaceutical Industries, the world's biggest maker of generic drugs. Teva, whose market capitalization is traditionally the biggest on the TASE, stands apart from most of Israeli business sector by virtue of the fact that it has no controlling shareholder, belongs to no business group, and is a rare instance of Israeli company enjoying the status of leading player in a global industry. Another important feature of the business groups was that four-fifths of those traded on the TASE in 2010 were structured as pyramids, in which a single company at the top controls successive tiers of increasingly larger numbers of subsidiaries and affiliates further down in the corporate structure. The system preserves management control at the lower reaches of the pyramid while minimizing for the controlling entity to have a proportionate amount of capital invested and at risk. Not counting Teva, only 12% of TASE companies in 2010 had a dispersed ownership (i.e., no single controlling shareholder) and half of those companies were dual-listed in foreign exchanges.[6]

Israel is not the only economy where business groups enjoy such an overwhelming presence, but it is unusual in the extent to which they have been so dominant and the fact that they have remained so long after Israel developed a relatively sophisticated and well-regulated capital market. Of 23 countries surveyed by a government committee examining the holding-group phenomenon, only Thailand, the Philippines, and Indonesia exceeded Israel in the percentage of local market capitalization belonging to the country's top 10 business groups. Among OECD countries, the group of countries against which Israel is typically measured, no other economy was so dominated by business groups.[7]

The business elites that control these groups have not been static, but the phenomenon has shown a remarkable persistence in the face of changing economic conditions, increasingly hostile regulations, and shifting ownership. Their origins of the business groups pre-date the founding of Israel in 1948 when the institutions of the pre-state Jewish settlement—the Jewish Agency and Histadrut labor federation—formed businesses and banks to aid in developing the economy and creating employment. These business groups persisted after independence, growing and expanding and joined by private-sector groups, formed with the encouragement of the government as a third channel for attracting investment from overseas. In an era of pervasive government control over the economy and finance, through an array of state, quasi-state, labor union, and private-sector companies, the holding groups effectively served as a conduit for implementing government development plans, channeling capital and creating the defense and other industries essentially by government fiat.[8] The era of government micromanagement of the economy came to abrupt end in the 1980s and early 1990s as slow economic growth and hyperinflation forced the government to change policy. Left to their own, heavily indebted and encumbered by an array of holdings that were unprofitable and poorly managed, the Histadrut and Jewish Agency were both forced to rapidly divest their business groups. The government sold off most of its businesses as well, as part of the privatization process aimed at making the economy more efficient. Thus, the Histadrut, which controlled 23% of net national product at the peak of its power in the 1960s and 1970s, saw its share fall to 14% at the start of the 1990s.[9] Today, its presence in the business sector is almost nil. The government's share declined from 27% in 1985 to 6% a decade later.[10]

But rather than leading to their dissolution, privatization and divestment gave new impetus to the phenomenon as the companies put up for sale were largely acquired by veteran or emergent business groups. Koor, the Histadrut's industrial holding company, was acquired by the Canadian Bronfman family and eventually sold to the IDB group, which had been created by the Recanati family and early in the 2000s sold to a shareholder group led by Noch Dankner, a scion of one of Israel's wealthiest families. The Israel Corp., a private-sector holding group, acquired state-owned Israel Chemicals, Oil Refineries Ltd. and Zim. The 1990s transitions brought to the fore among the business groups new and for most part self-made entrepreneurs, such as Lev Leviev, an immigrant from the former Soviet Union who made his first fortune trading in diamonds before acquiring Africa Israel Investments; Yizthak Tshuva, who leveraged a small contracting business to take control of Delek Group; and the Ofer family which used its shipping fortune to take control of The Israel Corp.

In more recent years, especially after the outbreak of the 2011 social justice protests, the government has sought to whittle away at the big holding groups. Among other things, the issuance of dual shares (with differing capital and voting rights) has been banned and the use of surplus voting rights as a means of retaining control has been constrained since the 1990s. Banks and other financial-service companies have had ceilings placed on their non-financial holdings. Most important of all was passage of the Economic Concentration Law at the end of 2013, which strikes at the heart of the holding-group phenomenon. Most significantly, it bars publicly traded groups structured as pyramids to reduce the number of tiers of companies to no more than three after four years and to just two after six years.[11] The law also prohibits cross-holdings between large non-financial and financial entities, defined as those having more than 40 billion shekels in assets, and prohibits financial entities from holding more than 10% of a non-financial entity. In addition, boards must comprise a majority of independent directors, except where a special waiver is obtained. The law's impact will only become evident in the next decade because the deadlines for meeting its most important terms extend for as long as six years. Meantime, however, many of the groups in the immediate aftermath of the law's passage began acting to confirm to its provisions.

In addition, the global financial crisis had deleterious effect on many of the groups. Although they traditionally showed a strong preference for operating in the domestic economy, the property booms in the USA and Europe in the years before 2008 encouraged many of the groups to invest heavily in overseas real estate. The property bust and the recession that set in 2008 forced the groups to write down and divest assets at a loss while the seizing up of the global financial markets and their heavy leveraging prevented them from rolling over debt. The retrenchment that followed has reduced the influence of the largest groups in recent years as they contend with debt or pare down their holdings to a core business. Many of the smaller groups have disappeared.

What has been the impact of these business groups? The Committee for Increasing Competitiveness in the Economy, which explored the issue in 2011, offered several factors that bear on the economy's performance.

The first relates to the internal capital market created by large business groups, particularly those structured as pyramids, whereby the controlling shareholder or shareholders tend to favor investment inside the group in order to retain more of the profits and control, even if alternatives with outsider partners would be preferable from the point of view of the affiliated company. These structures tend to create an imbalance between profit and risk, leaving the controlling shareholder with an outsized proportion of profit while saddling a disproportionate amount of the risk with minority shareholders and lenders. On the level of the broad economy, this risked distorting the allocations of capital, returns on investment and ultimately to lower economic growth.[12] In Israel, there is strong indirect evidence that these kinds of perverse incentives play a role in the persistence of business groups by virtue of the fact that shareholders are prepared to pay a considerable premium for control of a company. That premium—the amount that a buyer is usually willing to pay over the current market price to gain control over a publicly traded company—was 27% in Israel, a rate almost twice the average of 39 countries surveyed by the committee and nearly three times the average OECD level.[13]

The second relates to the quality and priorities of management on the companies belonging to business groups. A controlling shareholder whose principal interest is ensuring that profits accrue to him further up the pyramid is likely to choose managers based on personal loyalty, rather than on their ability to enhance the company's value for all shareholders.[14] There is evidence to suggest that in the fact that TASE investors have tended to assign lower values to companies affiliated with business groups. Konstantin Kosenko, in a 2007 paper for the Bank of Israel on business concentration, demonstrated that by comparing the Tobin's Q (the ratio of a company's market capitalization versus the value of its underlying assets) for companies affiliated with a business group and those that were not. What he found was the first group's Tobin's Q was lower (1.15) than for others (1.24), suggesting that investors assigned a discount to companies affiliated with a business group,[15] a telling statistic in light of the premium controlling shareholders were willing to pay for the same companies.

Finally, there is the deleterious effect the business groups have had on competition. In fact, it is rare that any one group has a monopoly presence in a particular domestic market, but it is very likely that in a particular segment (such as banking, food retailing, and telecommunications) two or three groups will share a market. The problem is exacerbated by the groups' presence in multiple specific markets, which creates a situation where group A will avoid entering a sector where group B is dominant for fear of a competitive counter-assault in a sector where group A is dominant. As a result, the affiliated companies of business groups tend to focus on markets where competition is minimal and the players are entrenched.[16] Thus, until their hold on it was broken in the middle of 2012, the Israeli mobile telephony market was effectively controlled by three companies, each affiliated with a business group (Cellcom by the Dankner family's IDB Group, Partner by Ilan Ben-Dov's Scailex and Pelephone by Shaul Elovich's Eurocom Group). When changing regulations introduced more effective competition to the market, the new players emerged mainly from outside the business groups, with the two major entrants controlled by French entrepreneurs.[17] Although companies affiliated with business groups are generally more profitable than their non-affiliated piers on the TASE, they spend less on research and development, take on more debt as a portion of equity, and grow more slowly.[18]

Kosenko concluded that the business groups had no advantage over the financial markets in allocating resources. The predominance of the groups makes control over much of Israel business sector dependent on personal relationships within the cadre of controlling families and the "strategies and preferences of a limited number of individuals." Their raison d'être was tied up in "prestige, political ties, family considerations and other factor than economic efficiency."[19]

One indicator of the extent to which they exploit their domestic market power as against deploying knowledge or other resources is the extent to which they have established an export presence or expanded overseas—two obvious outlets for companies seeking to grow in a small and generally mature market. Among four of the biggest groups—IDB, The Israel Corp., Africa Israel Investments, and Delek Group—the preponderance of their holdings is geared toward companies focused on the domestic market, with an emphasis on finance, services, retail, and energy. All of them have companies operating overseas or geared toward exports, but they tend to be relatively small and often held as minority stakes. None of them have extensive interests in high technology, which is a purely global business. Where their overseas holdings are extensive, the groups tended, until the collapse of global financial markets in 2008, to be focused in real estate where as it turns out they showed no particular insight or skills and sustained big losses.

The large business groups have drawn the lion's share of attention, but there are also issues about the absence of competition in Israel's domestic market for products and services. Although Israel has undergone a far-reaching process of trade liberalization and competition, laws have been toughened over the years, and the economy remains highly taxed and highly regulated. Moreover, those efforts at liberalization have run up against the inevitable inefficiencies of a small economy. The small size of the consumer market, as well as the entrenched position of existing players, makes it of little interest to foreign companies and raises barriers to new domestic players. The economy's comparative isolation—geographically distant from other developed economies and belonging to no trade bloc—magnifies that challenges it faces creating competitive markets. Thus, no foreign bank has ever established a retail operation in Israel (although they do engage in corporate lending) which has left banking in the hands of five major banking groups. Although three of the biggest companies are foreign owned, the food industry is dominated by five domestic manufacturers. The cellular telephony sector, long in the hands of three major players, underwent a government-imposed process of enhanced completion in 2012, with the introduction of eased rules for consumers changing providers and the introduction of mobile virtual network operators (MVNOs) that increased the number of competitors to eight. But it is doubtful whether a fully saturated market of Israel's size can sustain so many operators profitably over time. As of mid-2015, two of three MVNOs left the market while others are struggling with sharply diminished profits or even losses. Even as usage had grown with the widespread use of smartphones, the key industry benchmark of average revenue per user in Israel in 2014 was just \$74, a drop from \$108 three years earlier that reflected the new era of price competition. But ARPU was just a little over half the \$139 it was in similar markets overseas in 2014, which suggests that the decline was excessive. Meanwhile, the rate of earnings before interest, taxes, depreciation, and amortization for the three veteran cellular companies was 26% versus

29% in 2014 and had fallen even further for the Israeli operators in the first half of 2015 to about 18%.[20]

Although each product and service market in Israel has its own special characteristics, the food industry captures many of the issues and challenges facing the country's non-knowledge economy. The industry is needless to say quite large in terms of the overall economy, with turnover reaching 57 billion shekels ($15 billion) in 2009, equal to about 16% of total industrial turnover,[21] while the food retail industry had a turnover of another 55 billion shekels.[22] Food manufacturing employed about 58,000 people and the retailing side another 16,000, all told equal to about 45% the number employed in high technology, not counting the defense industry. In manufacturing, however, they earned at an average salary of 46 shekels an hour or $12.50 at the 2009 exchange rate. Concerns about the high retail price of food and the factors behind it (cottage cheese, serving as clarion call) sparked the social justice protests in the summer of 2011 and led the government to appoint a committee to examine the industry and make recommendations. From the interim report of the panel (popularly known as the Kedmi committee), three characteristics stand out—that industry is highly concentrated among a few manufacturers and retailers, import competition is restricted, and prices to the consumer are high by international standards.

Despite the presence of some 1700 manufacturers in the sector, some 40% of sales are in the hands of four groups—Tnuva, Strauss, Osem, and Central Bottling. A similar level of concentration exists in retailing, and it has been growing: In 1999, the supermarket chains accounted for 47% of sales; by 2009, the proportion had grown to 60% at the expense of smaller retailers and open-air markets.[23] Among supermarket chains, discounters and others have begun to appear in recent years, but the two biggest retailers (Shufersal and Mega) dominated the sector, accounting for 59% of chain sales and 86% of floor space in 2010. No foreign retailer is in the Israeli market. In Britain, by comparison, in 2007, the two largest chains (Tesco and Asda, the latter owned by Walmart of the USA) controlled 39% of sales for big-format stores and next two largest (Sainsbury's and Morrisons) controlled 23.7%. Moreover, among mid-sized supermarkets in Britain, the list of competitors is wholly different than in the big-format segment, diluting the power of the biggest retailers.[24]

In Israel, the barriers to small and new players penetrating either the manufacturing or the retail market are high. Big manufacturers and big retailers have pricing power that they use to discriminate against smaller rivals. Smaller companies cannot afford the costs of meeting regulatory requirements, which involve an investment in time and resources, and inadvertently or not serve to protect entrenched players more than protecting consumer or ensuring fair competition. Smaller companies also have difficulty accessing credit of that kind that matches their needs. Given the competition for shelf space at retailers and the lack of pricing power on the part of smaller manufacturers, investment in marketing and branding is likely to prove futile.[25] In the manufacturing end, there has been a trend toward mergers and acquisitions, with bigger players swallowing up smaller ones. In the retail segment, the availability of real estate in prime retail areas is limited and already taken by veteran chains.[26] Indeed, the real estate barrier to competition points up the extent to which the absence of competition is interlocking, i.e., that the absence of competition in one sector limits competition in others, thus making the government's task of encouraging competition extremely difficult. In this case, it is a highly regulated land sector that exacerbates scarcity (due to the country's small size) with problems of monopoly control (the state is the dominant land owner). Thus, Rami Levy, the biggest of a clutch of up-and-coming discount retailers, took three decades from its founding in 1976 to grow to eight branches. The company succeeded in tripling the number of branches in the five years after going public on the TASE but as of the middle of 2015, they numbered only 33. By comparison, the two leading retailers had 281 (Shufersal) and more than 183 (Mega).[27]

Imports could provide a source of increased competition in the food sector, but the small size of the market together with regulatory and trade restrictions creates barriers to entry. Between 1991 and 2000, Israel reduced tariffs on processed food and other manufactured products to a maximum rate of 8% for intermediate goods and 12% for final goods.[28] Tariffs on processed food and agriculture, in particular, were lowered from 1996 to 2003. But the scope of the reductions was more limited than in other sectors, and tariffs remain relatively high, which is mainly due to efforts to protect domestic agriculture,[29] an industry in which government regulation is extensive and designed to protect growers rather than ensure competition or encourage low prices. Those barriers are magnified by the small size of the Israeli consumer market, which means that many products are imported by a single importer who has an exclusive contract with the foreign manufacturer and thus no effective competition for a particular brand. Importers can sometimes bring in the same products via third parties, but they often have trouble obtaining the proper documentation and getting their products onto the store shelves of the big retailers who don't want to offend the foreign manufacturers.[30] Food that is sold through the biggest retail networks has to meet the standards of Jewish religious law (*kashrut*), which creates another high, if not insurmountable, barrier to foreign products. The government's Standards Institute, which creates quality measures for locally sold products and enforces them, raises the barrier higher still by imposing requirements of food and other imports that are often unique to Israel while charging high fees for its mandatory services.[31]

Not surprisingly, food costs in Israel are relatively high vis-a-vis levels of income. On the basis of purchasing power parity, a Bank of Israel study found that food prices in Israel were 15% higher in 2008 than the OECD average, although Israel's per capita income is below the average for the group. That means prices were, in fact, about a fifth higher than expected given Israel's income level.[32] In particular, prices were high for daily products, fish, and soft drinks, three categories where foreign competition for various reasons is particularly restricted. While milk is subsidized in many OECD economies, most notably in nearby Europe, Israeli tariffs on milk are 150%, preventing consumers from taking advantage of EU subsidies and effectively making Israel a closed market for dairy products.[33] The government has taken steps since the 2011 social justice protests to introduce more competition into the food sector, most notably with the 2015 Food Law as well as through stepped-up price controls, easing of approvals to import some categories of food products and lower customs. A 2015 study showed food price rises moderated after 2011 and even fell in 2014 and early 2015, but the decline was a relatively modest 1.3%, which left them 13.4% higher after overall inflation over the last decade.[34]

Interestingly, among the leading companies in the food sector, which is a typical inward-focused industry, Strauss Group in many ways defies the rule. Although the Strauss family holds a controlling stake and is often counted among the country's tycoon class, the family is focused on the single industry of food manufacturing rather than on building a pyramid-structured holding group. They started with a family dairy, gradually expanding into other segments like prepared salads and finally acquiring Elite Group, Israel's biggest maker of coffee and confectionaries, in 2004. Under its previous owners, Elite had stumbled badly in an overseas expansion drive in its coffee business in the 1990s. Under the control of the Strauss family, however, the group has embarked on a more successful and diversified expansion in a rare instance of an Israeli company leveraging knowledge assets in an old-economy industry not only in foreign markets but with new and innovative twists on traditional products, entailing branding and aimed at the world's most sophisticated consumers.

In 2014, some 55% of its revenue came from international activities.[35] Most of that was in the coffee business, where Strauss is a leading player in the Central and Eastern European, and Brazilian markets for roast and ground coffee and related products and services. The company also has a joint venture with PepsiCo of the USA to make and market hummus (the chickpea spread that can fairly lay claim to being Israel's national dish) and other prepared salads and dips under the Sabra brand in the USA and more recently in other global markets. The joint venture controls more than half of the American market and has played a leading role in turning hummus from a small ethnic category into a mainstream food item.[36] Strauss also has a water filtration and purification business, launched in 2006, based on proprietary technology developed in house and acquired from another Israeli company, and operates in China and Britain through joint venture with local partners (Haier Group and Virgin Group, respectively). Finally, it is developing a chain of high-end chocolate bars operating under the Max Brenner name. Max Brenner is a tiny business (as of the end of 2014, it had 58 outlets) but it operates in five countries and represents a wholly new business concept in the retail sector.

Strauss is an exception to the rule that outside of the technology sector Israeli companies very rarely use their strong position in the domestic market to venture overseas. With their limited domestic base, companies based in small countries like Israel are naturally at a disadvantage to rivals in bigger countries in penetrating the global market, but there are enough exceptions that it shouldn't be regarded a foregone conclusion. Switzerland counts both specialist consumer brands such as Rolex, the Swatch Group, and Lindt as well as mass-market companies such as Nestle. It has a world-leading pharmaceutical industry that includes Novartis and Roche as well as a major global presence in banking (UBS and Credit Suisse) and insurance (Zurich Insurance Group and Swiss Re). Sweden numbers technology companies like Ericsson, makers of consumer goods (Electrolux), global retailers (Ikea, H&M), pharmaceuticals (AstraZeneca, a Swedish-British group) and engineering (ABB, a Swedish-Swiss group). Both countries are inside Europe, which gives them an advantage of proximity to large markets, but Switzerland does not belong to the European Union and neither country belongs to the euro zone. In any case, Singapore, a tiny Asian country like Israel also distant from major markets, counts electronics manufacturing (Flextronics), Singapore Telecommunications, and Singapore Airlines among major multinational corporations.

Measured by representation on Forbes list of the 2000 biggest publicly traded companies by market capitalization, Israel counts a respectable 10. However, only two of the 10 can be regarded as global businesses (Teva Pharmaceuticals and Check Point Software Technologies), which suggests that for the other eight, their size is a function of limited competition in a relatively small domestic market.[37] By comparison, Sweden counts 24 companies in the top 2000, Switzerland 46, Singapore and Finland 20 each, and Ireland 11 (not counting foreign companies domiciled in the country for tax purposes). There are many factors at work behind Israel's relative failure. Among them is the difficulty Israelis have in building and maintaining large enterprises, a phenomenon discussed in more depth in Chapter 8. But without a doubt, the business-environment dissonance between a domestic market of cartels and heavy regulation and a global economy where competition is more intense and the ability to influence regulation virtually nil is a factor as well. Ironically, this creates a situation in which the Israeli firms with the heft to go global are the ones least prepared to succeed at it.

The businesses in Israel that do compete globally are often the creation of a single individual with a particular drive and vision rather than because of a favorable environment of economic fundamentals. Eli Hurwitz in the role of chief executive officer made Teva the world's biggest maker of generic drugs. The company floundered after he stepped down. Stef Wertheimer did the same establishing the machine tools maker Iscar that was eventually bought to American investor Warren Buffet while Morris Kahn leveraged a local Yellow Pages business into Amdocs, the dominant player in telecoms billing and services, and Gil Schwed turned Check Point into the leading maker of computer-network firewalls. But these were all standalone businesses unaffiliated with Israel's business groups, whose record at extending their reach outside of their home base has been poor. The knowledge and human assets they have acquired in their home market are inadequate or irrelevant abroad. Thus, with large swathes of Israeli business by inclination and strategy confined to the domestic market, exports amounted to 30.7% of the country's gross domestic product in 2015, far less than other small, advanced economies—45.6% for Sweden, 62.5% for Switzerland and (due to reexports) 124% for Ireland, and 176.5% for Singapore.[38]

That said, the Israeli companies that do export are highly competitive in terms of labor productivity. While overall, Israeli labor productivity was 14% lower than the OECD average per employee, there were sharp differentials between the country's export-oriented sectors and those geared to the domestic market, a characteristic Israel shares only with Germany and Denmark among OECD countries. The Israeli productivity advantage over the OECD average was in the tech-heavy, export-oriented segments of medical equipment, optical equipment, and metrology equipment sector was 40%.[39] Israel had an advantage in the electronics and chemicals sectors as well, albeit smaller. But this competitiveness is restricted to a relatively narrow segment of the Israeli economy as evidenced by the concentration of overseas assets and exports in a small group of companies. Among the top 20 Israeli multinationals in terms of assets held abroad, a good indicator of a company's global profile, a single company (Teva) accounted for the lion's share in terms of sales or 44% of the $35.1 billion total in 2010. Three others companies—Israel Chemicals via its parent The Israel Corp., Amdocs, and Makhteshim Agan Industries (since sold to a Chinese company and renamed Adama)—accounted for another 26%.[40] In terms of assets, the top four companies accounted for 50% of all foreign assets among the top 20.[41] Meanwhile, the country's 10 biggest exporters account for a rapidly growing proportion of total exports—increasing from 36.5% in 2007 to 51.3% in 2015.[42] The biggest of all is Intel, the US company whose semiconductor plant in Kiryat Gat sold $4.25 billion worth of products overseas in 2014, equal to close to 6% of Israel's total merchandise exports of $77.5 billion.[43] Among the other nine, five were in the chemicals or oil-refining sector (Teva, Israel Chemicals, Adama, Oil Refineries Ltd., and Paz), two are defense companies (Elbit Systems Israel Aerospace Industries), one is a maker of machine tools (Iscar) and one in high-tech (HP Indigo, a unit of America's HP). Interestingly, while the two defense contractors specialize in defense electronics, the list of top 10 exporters does not include any pure high-tech companies except the two Israeli units of US firms.

In spite of the country's reputation for innovation and advanced technology, most Israeli business are not in the global forefront of innovation either by introducing new products or services or by utilizing the most advanced organization or managerial structures. Relative to other countries belonging to the OECD, Israeli companies show relatively high levels of product and process innovations, according to OECD data. More than 80% of Israel's large enterprises and more than half of small- and medium-sized companies reported doing some kind of innovative activity, ranking it sixth among 34 countries surveyed.[44] But the data are misleading: The Israeli figures are for the years 2006–2008, while most of the other countries reporting did so for 2008–2010, years of deep recession for most OECD countries and coincident with a sharp drop in business R&D spending in 2006–2009 across the rest of the OECD.[45] In Israel, meanwhile, there was no slowdown to speak of all in those years, yet the percentage of Israeli companies saying that had introduced technological innovations, such as new or significantly improved products, adopting new manufacturing processes, adopting innovative marketing methods, or undertaking organizational changes, declined over the relevant years. Among 2316 companies surveyed by the Central Bureau of Statistics for 2010–2012, 25% reported they had introduced a process or product innovation during the period, down eight percentage points from the previous (and first) survey covering 2006–2008.[46] Even among big companies, which are responsible for the great majority of innovation, the number reporting significant innovation in the period dropped six points to 62%.[47] Another telling statistic in this regard is Israel's international trade in knowledge assets, which encompasses such things as patents and licenses, transfer of trademarks and patents, and industrial R&D. As an economy heavily slanted toward high-tech R&D, Israel should show a heavy weighting to receipts as against payments, but the balance is unusually lopsided: Receipts of knowledge assets in 2012 amounted to 5.15% of GDP, the fourth-highest among 32 OECD countries surveyed, but its payments were just 0.95% of GDP, putting it in the mid-range.[48] No country has a gap anywhere approaching Israel's, suggesting the country is not making use of foreign innovation nearly as much as it could.

Israeli adoption of cloud computing serves as a good barometer for the phenomenon of lagging innovation in Israeli business. A survey by the global consulting firm Gartner found that in the USA, 10% of all software spending goes to cloud software while a parallel survey by the Israeli research company STKI found it was only 3%.[49] In contrast to its high levels of R&D spending, Israeli spending on information technology is relatively low by developed-country standards. A Gartner survey from 2011 put it at 3.1% of GDP, versus a range of 3.5–5% for Western European countries and 6% for the USA.

In any case, the innovation that exists in Israel is skewed toward the high-tech industry, as opposed to innovation in older, more mature industries, according to the CBS survey, with 59% of information and communication firms reporting that they had undertaken technological innovations during the three survey years. On the reverse side of the coin, not only did older, more traditional industries like food service and mining innovate less, but so did old-economy sectors where innovation would be normally be considered critical, such as financial services and, with the rapid development being undertaken in e-commerce, retailing. In the first three sectors, the percentage of companies reporting they had introduced some kind of technology innovation was about 18% and in the case of retail just 9%.[50] Given the great divide in Israel between the export and domestic sectors, that is not surprising: Financial services and retail are geared to the domestic market and largely sheltered from competition and the need to innovate. The divide is further backed up by the difference between companies that innovate and those that don't: Among innovators, 33.3% said a key goal was to develop markets overseas; among non-innovators, only 8.2% reported any interest in foreign markets.[51]

Among the barriers to innovation cited by companies in a CBS survey covering the years 2006–2008, 79% cited costs and the ability to finance, which might be expected in a small economy where smaller firms are the norm. But large proportions cited the absence of any need to innovate (33%) and the view that the market is controlled by other companies (32%), responses that reinforce the view that there is an absence of effective competition in large sectors of the economy.[52] While Israeli industry has seen a huge growth in the number of staff with at least a post-secondary-level education or advanced professional training that has not brought a commensurate improvement in innovative skills, an issue discussed in Chapter 7. The percentage of jobs among manufacturing companies designated for people with a higher education or specialized training nearly doubled to 63% in the 15 years to 2012,[53] but among companies polled for 2006–2008, 34% cited the lack of skilled and trained personnel to undertake innovations, a figure that points to serious human resource issues.

The problems that characterize the great majority of Israel's private sector companies are shared by the government, often in a more extreme form. Globally, the state's reputation is mainly informed by the stellar reputation of its army and intelligence services, which have earned Israel a remarkable ranking as the eighth most powerful country in the world by a US News & World Report survey,[54] despite a population in the single-digit millions. However, as discussed in more detail in Chapter 8, the army and intelligence services operate more like a start-up company geared toward a mission, an environment where Israelis excel, rather than as a large enterprise engaged in routine operations, where they do not. In times of war or in confronting terror, they rely on personal initiative, innovative thinking, and often by breaking the rules. But on a day-to-day basis, overall organization in the defense establishment is characterized by poor organization, rigid management, and waste. In the case of the Israel Defense Forces, whose operations are far more open to outside observers than the intelligence services, the innovation and advanced technology employed in purely military activities isn't applied to the day-to-day administration of the army. And that kind of administration is more typical of the other branches of government machinery that don't face the exigencies of war or terrorism. Although government spending as percentage of GDP has fallen, the civil service remains bureaucratic and inefficient and often corrupt, in the latter case certainly on the local level. Low scores for factors relating to government account for Israel's relatively poor ranking in the World Economic Forum's Global Competitiveness Index, where it placed 24th among 138 countries in 2016.[55] Israeli institutions ranked 31st and for goods market efficiency (a function mainly of regulation and bureaucracy) it ranked 32nd. On specific issues related to government, Israel scored particularly badly—46th for the burden of regulation, 78th for wasteful of government spending, 46th for favoritism by officials, and 39th for government transparency.[56] "Inefficient government bureaucracy" was cited by businesspeople surveyed by the WEF as the "most problematic factor for business."

The relative inefficiency of government compared to the private sector is illustrated by infrastructure. Overall, Israel ranked 25th in the WEF index in 2016 in that category, but that was due to relatively high rankings of 19 for cellular communications and airline-seat availability, both of which are controlled by the private sector and have been subjected to reforms in recent years creating highly competitive markets. However, where the government is directly involved in owning, developing, and operating infrastructure, Israel's rankings were much lower—46 for roads, 56 for ports, and 60 for airports.[57]

The weakness of Israel's government sector is traceable both to politics and to administrative culture. As noted in Chapter 3, Israeli governments tend to be relatively short-lived—between 1990 and 2010, the average coalition lasted 2.35 years, slightly below the 2.62 average for 27 parliament democracies surveyed by the Israel Democracy Institute.[58] There has also been a far more frequent change in the ruling party between governments in the 1990–2010 than there was in the past when the Labor Party dominated politics, so that among 27 democracies surveyed by IDI in that period Israel experienced more changes in ruling party (six in total) than all but two of them.[59] Given the overall economic consensus over those years, the frequent changes haven't so much led to policy zigzags, but they have led to frequent personnel changes in the Finance Ministry, which is by far the most powerful economic-policy making body in the government. The average term for a finance minister in Israel since the founding of the state was 836 days, putting it 14th among 19 surveyed democracies.[60] In fact, that overstates the real average, which was raised by the 11-year tenure of Levi Eshkol in the 1960s. In the post-1990 era, the average stint for an Israeli finance minister was 47% shorter than in the pre-1990 period. The arrival of a new finance minister doesn't necessarily spell an ideological break with his predecessor, but it does mean frequent running-in periods as the new officeholder learns the job and frequent policy reversals as he cancels his predecessor's programs and initiates new ones for which he can then take political credit. For same reason, the timeframe finance and other ministers can allow for policies to bear fruit is also short because they don't expect to remain in office to see the results and reap the political benefits. As a result, strategies for coping with problems such as rising home prices or lagging labor productivity that require time are often poorly conceived if addressed at all. Under the circumstances, it is difficult for the government to adhere and implement any consistent, long-term policies beyond the broad consensus over fiscal discipline.

That said, the weakness of the government sector is not just a function of policy but of the civil service and public sector generally. Unlike the private sector, where union membership has dropped precipitously since the 1980s, organized labor retains a tight grip on the civil service and in state-owned enterprises. Labor conditions are dictated by collective agreements and reforms subject to negotiations with unions, which have effectively blocked any efforts at instituting modern management practices into government. Among the problems cited in a 2013 report that recommended major changes to the civil service found the hiring process lengthy (an average of 74 days from the deadline for a job opening until a candidate was hired) and based on outdated criteria.[61] Employees aren't subject to real performance benchmarks and no pay or other incentives are offered for the best employees. Performance reviews, such as they are, are designed to ensure pay raises rather than measure achievement, with an average score of 9.62 being awarded for civil servants on a scale of 1–10.[62] In its use and deployment of technology and e-government, the Israeli public sector resembles the broad economy more than it does the high-tech sector. The United Nations' E-Government Development Index in 2016 ranked Israel 20th overall and 18th for online services, which puts it at the low end of the world's most developed economies.[63] A digital initiative launched by the government in 2013 sought to address the deficiencies but it was not until 2017 that government bodies were required to something as basic as accept e-mail communications from individuals or businesses.[64] Much information that should been in the public domain and available for free, such as data on crime, weather and real estate transactions or the national geographic survey, were not. Different government bodies use different platforms and responsibility for digital accessibility is divided among different bodies.

Maritime ports, where Israel had a low 56th ranking in the WEF index, serve as a good illustration of the government sector's weaknesses. With 98% of all foreign trade going through the ports due to the absence of any significant land-transportation links to neighboring countries, Israel's ports are critical to the country's trade-dependent economy. Officially, they are a government monopoly but effective control lies with the labor unions that wield immense power by their ability to call strikes and labor slowdowns that have the potential of shutting down the economy. The unions fortify their grip at the political echelon by influencing internal primaries in the key political parties. There is some disagreement about how efficient Israeli ports are by global standards, but the frequent strikes and labor actions without a doubt impose direct costs to the economy estimated at about 400 million shekels annually.[65] High port fees due to wages that are three times the national average and to extensive corruption are passed on to businesses and consumers. The government's attempts to reform the system over the years repeatedly failed because of the power of the unions, so that in 2014, it opted to circumvent them by allowing privately owned ports to compete with the state-owned facilities.

The government succeeded in breaking the power of the unions in telecommunications and eventually in the ports, but on the balance its record has been poor and union-controlled monopolies still exist not only in the ports but also in electric power, water, and rail transportation.

It could be seen as symbolic that Israel's ports typify many of the inefficiencies of the broader economy: The human capital assets Israel has in the form of an elite of highly trained and technology-savvy engineers and entrepreneurs is in the main employed in the high-technology sector, a global business that has little to do with the surrounding economy and does its businesses over the Internet and by air travel because it deals with intellectual property more than material goods. For the rest of the economy, the ports constitute the final obstacle in a series that begins with a workforce with less than adequate skills and training, a business sector dominated by holding groups and cartels and a regulatory environment that stifles entrepreneurship and competition. The result is, in effect, two Israeli economies—one global, innovative, and competitive and the other confined to the domestic market with all the constraints imposed on it.

Notes

1. Central Bureau of Statistics, "ICT Sector Estimate for 2014" (Hebrew), August 12, 2015, 1.
2. Finance Ministry, *Weekly Economic Survey* (Hebrew), February 14, 2016, 6–8.
3. Bank of Israel, *Recent Economic Developments*, No. 131, May–August 2011.
4. Hagai Amit, "Mortgaging Their Future," *Ha'aretz*, September 29, 2012.
5. Committee for Increasing Competitiveness in the Economy (Interim report), September 19, 2011, 80.
6. Ibid., 79.
7. Ibid., 88.
8. Konstantin Kosenko, "Evolution of Private Business Groups in Israel," *Israel Economic Review*, Vol. 5, No. 2 (2007), 67.
9. Ibid., 68.
10. Ibid., 67.
11. Freshfields Bruckhaus Deringer LLP, "Israel Passes Landmark Legislation Aimed at Breaking Up Conglomerates," December 2013.
12. Committee for Increasing Competitiveness, 75.
13. Ibid., 88–89.
14. Ibid., 76.
15. Kosenko, 82.

16. Committee for Increasing Competitiveness, 77.
17. Of the other three, two were affiliated with groups (Alon Blue Square and Fishman Holdings), but they were offshoots of existing retail operations, entailing few costs and limited goals for growth or market share.
18. Kosenko, 82.
19. Ibid., 57.
20. Gad Peretz, "Consulting Firm: 'Cellular Rates in Israel Likely to Rise'" (Hebrew), *Globes*, August 30, 2015.
21. Interministerial Committee Examining Competition and Prices in the Food and Consumer Product Sector, Interim Report (Hebrew), November 11, 1, 17.
22. Ibid., 19.
23. Ibid., 19. More recent data from the retail consulting firm Czamanski & Ben Shahar show market share of the two largest supermarket chains falling. Since the food competition report was published, Mega went bankrupt and sold most of its assets to a medium-sized chain called Yenot Bitain. As a result, the top two chains now account for 43.5% of all supermarket sales. However, the lower market share for the big chain seems to have been the short-term result of an industry shakeup, and more recently there have been signs of increasing consolidation.
24. Ibid., 20.
25. Ibid., 23–24.
26. Ibid., 22–23.
27. Various companies' annual reports, first quarter of 2015.
28. Ibid., 27. More recently, the government has acted to ease these barriers by allowing what are known as "parallel imports," i.e., allowing multiple firms to import the same product. The effectiveness of this is yet to be seen.
29. Ibid., 28.
30. Ibid., 27.
31. Ibid., 25.
32. Bank of Israel, *Recent Economic Developments*, No. 132, September–December 2011, 36. The bank estimates given trends in inflation, the exchange rate and income, that the difference widened by another 10% in three years after 2008.
33. Ibid., 36.
34. Knesset Research and Information Center, *Follow-Up on the Implementation of the Law for Promoting Competition in the Food Sector, 2014* (Itamar Milrad) (Hebrew) (Jerusalem, June 8, 2015), 1. It should be noted that the period of price declines coincided with a sharp decline in world commodities prices.
35. Strauss Group, *2014 Annual Financial Statement*, 11.
36. David Kesmodel and Owen Fletcher, "Hummus Is Conquering America," *The Wall Street Journal*, April 30, 2013.

37. Chen Liyan and Andrea Murphy, "The World's Biggest Public Companies," *Forbes*, May 6, 2015.
38. World Bank, Exports of Goods and Services (% of GDP) (database).
39. Bank of Israel, *Annual Report 2013* (Jerusalem, February 4, 2014), 62–63.
40. Vale Columbia Center on Sustainable International Investment, Emerging Market Global Players Project, "Israeli Multinationals Back on Track After a Difficult Year." The list does not include financial companies.
41. Ibid., 3.
42. Israel Export Institute, Economics Department, "Developments and Trends in Israeli Exports," March 2016, 39.
43. Yuval Azoulay, "Intel Exports in 2015: $4.1 Billion," *Globes*, January 30, 2016.
44. OECD, *Science, Technology and Industry Scoreboard 2013* (Paris, October 23, 2013), 182.
45. Ibid., 26.
46. Central Bureau of Statistics, "Findings from the Business Innovation Survey 2010–2012" (Hebrew), July 10, 2014, 2. The survey encompassed firms employing at least 10 people (except for the computer services and R&D sectors, where the minimum was five) in industry, utilities, retail and wholesale services, hotels and restaurants, transportation, storage/logistics and communications, banking and finance and business services.
47. Ibid., 3.
48. OECD, *Science, Technology and Industry Scoreboard 2015* (Paris, October 19, 2015), 143.
49. Orr Hirschauge, "Israeli IT People Are Being Left Behind: Advances in Technology but Not in Business" (Hebrew), *TheMarker*, May 7, 2013. This is the latest survey available at the time of writing.
50. Ibid., 3.
51. "Business Innovation Survey 2010–2012." Table with breakdowns provided privately at the request of the author.
52. "Business Innovation Survey 2006–2008," 4. The more recent survey didn't ask these questions, so more recent data aren't available.
53. "Business Innovation Survey 2010–2012," 8.
54. "U.S. News & World Report, 2017 Best Countries," undated. The ranking was based almost wholly on a high score for "strong military." Other areas, such as leadership and political and economic influence, were scored much lower.
55. Klaus Schwab (ed.), *Global Competitiveness Report, 2016* (Geneva, 2016), 210. Most of the rankings are based on survey questions and the rest on objective factors.

56. Ibid., 211.
57. Ibid.
58. Shlomit Barnea, Shurik Dryshpitz, and Ofer Kenig, "Government Stability—Working Paper," undated, 14.
59. Ibid., 19–20.
60. Ibid., 30.
61. *Report of the Reform Committee for Improving Human Capital Management in the Public Sector* (Hebrew), June 25, 2013, 83–99.
62. Meirav Arlosoroff, "Civil Service Reform Means Getting Bureaucrats to Work," *Ha'aretz*, July 3, 2013.
63. *United Nations E-Government Survey 2016* (New York, 2016), 111 and 83.
64. Tehilla Schwartz Altshuler, "Time for the Start-Up Nation to Building a Start-Up Government," March 20, 2017.
65. Olaf Merk and Thai Thanh Dang, *Efficiency of World Ports in Container and Bulk Cargo (Oil, Coal, Ores and Grain)* (Paris, 2012), 10; Yona Rubinstein, *School of Economics Policy Paper: Should Additional Seaports Be Developed and if so Under Government or Private Ownership?* (Herzilya, June 2014), 10–11; and *Report of the Committee for Socio-Economic Change* (Trajtenberg committee), 176.

CHAPTER 5

The Enterprise as Product

Israel's high-technology industry is an industry of start-ups—companies that are rarely more than a few years old, employ few people, devote their human and financial resources to research and development, managed by their founders and privately owned. In other words, it is very much an industry of entrepreneurship—the product of the personal initiative, skills, and vision of the founder or founders and guided and managed by them, more often than not. The Start-up Ecosystem Report for 2015 which surveyed companies in 20 technology clusters ranked Tel Aviv fifth in the world for its entrepreneurship activity, making it the highest ranked cluster outside the USA.[1] In the 10 years from 2006 to 2015, the Israeli high-tech industry has come close to doubling the number start-ups formed annually from a range of 550–650 a year in 2006–2009 to between 1050 and 1150 in 2011–2014.[2] The figures that almost certainly understate the extent of the phenomenon since many start-ups don't raise money from outside investors or enroll in government programs and therefore aren't captured in the data. Moreover, the start-up phenomenon has shown itself remarkably persistent in the face of global recession, the cyclical nature of the venture capital industry, and Israel's volatile geopolitical situation. At the depth of the global recession in 2008–2009, which was accompanied by a sharp drop in VC financing,[3] the number of new start-ups formed declined to 580–600 a year, but that was only fractionally down from their peak of 668 in 2007, and the number that closed remained virtually unchanged. After a sharp drop in 2010, the number of new-company formations surged in 2011

D. Rosenberg, *Israel's Technology Economy*, Middle East in Focus,
https://doi.org/10.1007/978-3-319-76654-6_5

to 1100 and remained at that level for each of the next four years even though the seed-stage investment on and exit activity—two key factors in start-up's prospects for success—were both stagnant.

All of this suggests that the start-up phenomenon in Israel isn't fully governed by the usual considerations that go into setting up a new business, such as a ready and available market, business confidence, favorable industry regulations, and the availability of finance. Rather, it has become something akin to a family business wrought on a national scale: Israelis start up companies because it is the regarded as the natural and obvious thing to do for those with the relevant skill sets, and educational and/or army background. The phenomenon is facilitated by an institutional infrastructure that begins with official government encouragement in the form of tax incentives, state-subsidized venture capital, a network of incubators (that have since been privatized), and research and development aid. An even more critical part of that infrastructure is a highly developed venture capital industry, which has been complemented more recently by an increasingly sophisticated community of angel investors and growing corporate investment, mainly by multinationals. It has also been aided by the presence of investment banks ready to shepherd young companies into initial public offerings in the USA and Europe; more recently, the Tel Aviv Stock Exchange itself has become more hospitable to young technology companies, especially in biotech and medical electronics. A third element is the plethora of multinational R&D firms operating in Israel that provide a training ground for future entrepreneurs and more recently financing and even office space for new enterprises. Israel's universities have a long history of commercializing innovations developed in their laboratories, thereby serving as an important source of start-ups. But none of these structures would have emerged on such as wide scale without the qualities Israelis themselves bring to the start-up phenomenon—a culture of personal initiative that values risk-taking and resists the hierarchy and discipline of large organizations and a worldview informed by the precariousness of the situation at any moment, all of which is discussed more fully in Chapter 8.

Still, as much as this mixture of infrastructure and culture creates an environment unusually friendly to start-ups, it acts just as much to prevent the emergence of larger enterprises capable of sustaining growth with a strategy of sustained and disciplined innovation, and the depth of management to operate them. Israel's high-technology sector is an industry of pure innovation: The "products" it creates are the enterprises themselves and the intellectual property they develop, which is then sold far more often than not in a mergers and acquisitions transaction. The Israeli technology industry's most successful product has been the start-up enterprise itself.

With few natural resources, a large domestic market or labor force, low costs or an environment of political certainty on which to build a globally competitive economy, Israel has done a remarkable job of harnessing the country's principal comparative advantage, namely its human capital, with its high-tech industry. Business sector research and development in 2013 accounted for 4.2% of the gross domestic product, the biggest among countries belonging to the Organization for Economic Cooperation and Development (OECD).[4] Among small, technology-focused economies, only Finland and Sweden (both 3.3%) came close. In the USA, the rate that year was 2.3%, and the OECD average was less than 2.4%. In Israel, that R&D investment is put to good use in the form of exports and in rewarding and well-paid employment, although its contribution should be kept in perspective. The information and communications technology (ICT) sector that forms the core of Israeli high technology accounted for 11% of the country's GDP in 2015 and a significantly bigger 19.2% proportion of its exports.[5] But after rapid growth in the first decades of the 2000s, ICT contribution to the economy has been flat. Growth in exports has been modest and much it due to the presence of a major Intel fabrication plant in Israel. Moreover, ICT's share of the country's total employment was less than 5% in 2015, although with average pre-tax salaries of about $70,000, about twice the national average, it makes up a much larger share of total compensation.[6] Moreover, Israel's technology balance of payments, which in many respects gives a better picture of the industry's strength than merchandise and services exports do because it measures the intangibles that are the start-up industry's focus, is deep in surplus. For the years 2007–2010, which included the global recession years of 2009–2010, technology exports ranged between just over $8 and $9.8 billion, far in excess of its technology imports.[7] As a proportion of GDP, Israel had the second biggest technology balance of payments surplus in the world after Ireland in 2008 (4.6% versus 14.3%) and far in excess of European economies with a technology orientation.[8] Finally, the start-up industry has attracted large amounts of foreign investment. In the 2006–2015 decade, Israeli technology companies raised approximately $22 billion from investors and generated more than $56 billion back in the form of exits.[9] On both sides of the equation, both in terms of capital invested in start-ups and the companies acquired through M&A deals or IPOs, the money is nearly all from overseas.

It is difficult to measure the contribution of start-up companies to economic output because many do not generate revenue at all or only at a fraction of their potential because they concentrate on R&D. But their value as potential generators of value-added in terms of the contribution of human capital can be roughly measured by the valuations placed on them by investors when they are successful enough to be acquired or conduct an initial public offering. As an example, seven start-up companies sold during 2008 had valuations that ranged between $1.9 and $11.3 million per employee.[10] None of them employed more than 150 people and one, Fraud Science, had just 15 people on its payroll. By comparison, among publicly traded companies on the Tel Aviv Stock Exchange, the market capitalization per employee for manufacturing and service companies was mainly in the range of $120,000 and $700,000,[11] but payrolls in these companies ranged from about 1500 to 38,000, which means their broader impact on the economy was greater.

The start-ups that comprise the core of Israel's high-tech industry have traditionally been highly focused on ICT, with a focus on business customers as against consumers. In the decade 2006–2015, life sciences, IT and enterprise software, and Internet start-ups captured nearly two-thirds of all venture capital investment in Israel, according to IVC Research Center data.[12] Other major segments included communications (17%), semiconductors (9%), and cleantech (4%). Until relatively recently, Israeli start-ups avoided consumer products and services, which would require market knowledge and savvy that Israeli companies have generally speaking failed at, as well as big and sustained capital investment. Nor have Israeli entrepreneurs entered large segments of the global technology industry, such as e-commerce, financial and healthcare services, and media or entertainment. Part of this is naturally due to the small size of the Israeli technology industry, whose start-ups raised $21.9 billion in venture capital in the 2006–2015 decade, compared with $333.5 billion by US start-ups in the same period.[13] The average deal size for an Israeli company during that period was $4.1 million, a little more than half the size in the USA.[14]

It is doubtful that lower costs play a factor in the relatively small capital requirement of Israeli start-ups, since Israeli engineers are not significantly less expensive than their American counterparts, nor are other overhead expenses. Indeed, many Israeli start-ups take on the significant added cost of setting up an office in the USA early in their development to access the American market, while a typical US enterprise at the same stage can work out a single location serving its home market. But even as the Israeli industry has matured, the preference for smaller start-ups raising relatively little capital hasn't changed. Averaging VC financing rounds for late-stage companies climbed after a 2009 trough to more than $21 million in 2015, but financing for seed, early and mid-stage companies, has been flat since 2006 and in the case of seed-stage companies was actually lower than its 2006 level in 2015.[15]

Another factor that affects the character of Israel high tech is the training, experience, and expertise of entrepreneurs themselves and their teams, which comes from the army and is thus focused on defense needs. Many start-up founders have employed their army training and experience directly in defense and homeland security applications, but the barriers to winning procurement contracts overseas are very high, both because governments are reluctant to source foreign technology and because Israel enforces a strict regime on exporting sensitive military and dual-use technology.[16] Thus, the defense roots of many Israeli entrepreneurs and engineers manifests itself in civilian applications of communications technology and network security, although it also shows up in unusual places such as robot vacuum cleaners. Even, the heavy weighting of high-tech investment in life sciences traces its origins not only to the universities, but also to the application of defense technology to medical electronics, for instance missile guidance technology used in medical diagnostic equipment.

However, the premier example of the influence of the defense background to Israel's high-tech industry is network security, which has emerged as a major sector globally as the risk of hacking attacks and other issues has grown in an increasingly interconnected world. A survey by the government's National Cyber Authority estimated that sales of so-called cyber technology reached $6 billion in 2014, exceeding Israel's conventional defense exports and comprised 10% of the world total.[17] Just as importantly, Israel was responsible for 15% of all cyber R&D worldwide, spending about $200 million in 2014, nearly four times what it spent four years earlier. The number of companies with commercial sales amounted to 300 in 2015, double the number four years earlier. Eight companies were acquired in M&A deal worth $700 million, but two others went public ensuring they would for now remain standalone businesses, including one (CyberArk) at a $2 billion valuation. The foundation of the cyber sector, which the government has made a national priority, helping to develop a cyber technology part in the southern city of Beersheva, is the army's intelligence units, most famously its 8200 unit (which is described in more detail in Chapter 8). But just as importantly is that Israel also counts some older, established network security companies, including Checkpoint, Verint, and Nice Systems. By one estimate, the veterans of army technology units like 8200 account for 20% of all the entrepreneurial talent behind Israeli cyber start-ups, with former employees of the veteran cyber companies accounting for the rest.[18]

An important phenomenon to recognize is that the ability to conceive and engineer new technology has proven in the Israeli context to be quite fungible. As detailed in Chapter 8, Israeli innovative capacities are less connected with knowledge and expertise that have acquired in a particular industry or science, such as defense communications, but in Israeli culture and the ability to apply digital knowledge to new applications. The rapid emergence of autonomous-car technology in Israel in recent years is a testament to that. Israel has never had an automobile industry, but the expertise required to develop self-driving cars and related technology doesn't require automotive-engineering skills but abilities in artificial intelligence, machine learning, mapping, image and video processing, 3D sensing, and even neuroscience. A lot of these technologies have their origins in military applications, which gives Israeli start-ups and the army background of many of their entrepreneurs and engineers leg up on the competition from other countries in the sector. Israeli start-ups to encompass basic self-driving technology (Mobileye, Valens), security for interconnected vehicles (Argus, Arilou), ride-sharing (Gett, Via, Moovit), and vehicle communications (Otonomo, Autotalks).[19] By one count, there were close to 150 automotive start-ups and multinational R&D centers operating in Israel in the segment, double the number in 2013 and the start-ups among them had raised in aggregate $820 million in 2014–2015 alone.[20]

Another example of the fungibility of technology is the consumer sector, where Israeli activity has grown in recent years as the cost of reaching the market has plummeted, and the required skills sets have changed. Thus, in the five years from 2006 to 2010, Internet companies increased their share of total venture capital fund-raising to 18% from 5%.[21] In fact, their share is likely to be higher because the unusually low start-up costs enable budding entrepreneurs to avoid the venture capital route altogether, either by bootstrapping (financing by the founders themselves) or relying on angel investors whose investments are less likely to be captured by the IVC Research Center figures. Israeli companies have overcome their comparative disadvantage in conventional marketing, which require a fine-tuned understanding of your target audiences cultures and values, with sophisticated tools for the measurement, collection, analysis, and reporting of data captured online. "Internet marketing is more analysis than plain marketing the way we understand it. It's not communication skills – what works best to make the user click," says Dror Nuhumi, a partner at the tech-investment fund Norwest Venture Partners. "In terms of communications skills, Israeli companies still suffer. It is more difficult to get a good presentation in front of my partnership, but when it comes to marketing analysis on the Internet, Israeli companies know how to do that very well. It is a way for them to be successful on Internet as well."[22]

Interestingly, the Israeli industry's skill set in this field was derived from online gambling, a market where several Israeli entrepreneurs proved successful with companies like Playtika and 888. Babylon, an online translation company, serves a good example of how these skills were transferred to other businesses. Its translation business was decimated by the advent of Google and other free software, and the company was sold in 2007 to Noam Lanir, who had acquired skills in Internet marketing and metrics by buying and developing online gambling companies in Britain. Under Lanir, Babylon began selling advertising space and services via Google and other search engines where users need translation tools to access online material. Using statistical tools that gave the company sophisticated analytical powers, Babylon used the exposure its translation software gave it to generate ad revenue and by 2012 was ranked 38th among the world's most frequently visited Web sites, with more than 90% of its revenue coming from advertising.[23] Babylon wasn't alone: A host of Israeli companies followed a similar strategy to become major players using a business model based on profit sharing with search engines to which they drove traffic. Indeed, so many Israeli companies were engaged in this business that they had become their own technology cluster known as "Download Valley." Based on controversial distribution methods and products of doubtful value to users, the business for Babylon and it peers ultimately proved to be short-lived,[24] but the experience was nevertheless indicative of the skills that would be applied elsewhere in reaching mass markets.

More recently, those skills have been parlayed into the emerging ad-tech industry, which develops technology, software, and services for delivering, controlling, and targeting online ads. In the first half of 2015, the Israeli sector counted more than 600 companies, a threefold increase in five years.[25] The sector employed 16,200 people in Israel and overseas and had combined revenues in 2014 of about $3 billion. Neither was the industry all start-ups: About half of the companies reported having revenues and the largest are industry leaders in their subsegments, including Outbrain and Taboola (content recommendation), IronSource (mobile and desktop applications distribution), SimilarWeb and Crossrider (Web site traffic monitoring and analysis), Matomy Media (online ad campaign management), and Kenshoo (marketing software). Although the sector was showing some signs of slowing growth in 2015 measured by the number of new start-ups formed, ad tech's share of total Israeli start-ups accounted for about 10% in 2014, up from 6.1% four years earlier. In 2014, ad-tech companies raised $511 million in the capital, up from $144 million two years earlier.

Life science companies have consistently accounted for a large segment of venture capital investment, accounting for more than a fifth of all VC investment in 2006–2015.[26] The number of companies in the field has grown from 467 in 2004 to 1380 in 2014, with an average of 98 new companies added every year over the decade[27] and annual fund-raising has grown nearly threefold to $930 million.[28] With its heavy burden of clinical trials and multiple national regulatory authorities to answer to, medical technology doesn't on the surface seem to be well suited to the Israeli preference for small, low-cost, and fast time-to-market technology, but in fact, a model has developed for drug developers that enables small R&D companies to complete the process by forming early collaboration with bigger partners that provide capital, regulatory experience, and marketing apparatus. "If you have new stent or *defibrillator*, you need to go through clinical stages, R&D, design and regulatory approvals," explains Jonathan Goldstein, who has worked as a vice president for development at various life science companies and is CEO of a Jerusalem-based start-up called Innovo Mimetics. "The cost of marketing today to the end user is so high that its likely business model is to go through one of the larger channels who are likely to [also] be your exit." Multinationals may collaborate on marketing with a start-up for a time, but ultimately they want to own the product. "It's difficult to find investment capital to go all the way to sale," he says. "Once you have gotten to scale you need a suitable partner, which increasingly is coming to the large corporates. If that's the case you may as well use the large corporate earlier on. It's difficult for start-up with a single product or even a platform of products to be self-sustaining."[29] In the years 2012–2014, life science companies enjoyed the biggest exits in the Israeli technology sector (with the exception of semiconductors whose average was inflated by some unusually large M&A deals in 2014) with an average of $164.3 million, compared with $145 million for IT/enterprise software companies, $141 million for communications start-ups, and $132.7 million for Internet companies.[30] Of these, two-thirds were sold in the research and development or initial revenues stage.[31] Relatively large numbers of biotech companies go public on the Tel Aviv Stock Exchange, but for all intents and purposes they remain start-ups, and their failure rate is more typical of the start-up industry than for publicly traded companies.[32]

Israeli tech entrepreneurs and venture capital funds that back them generally avoid cleantech. This is surprising in light of Israel's scientific achievements in water and solar technology, but the reason is the high and sustained start-up costs involved in environmental technology—regulatory approvals factor heavily and the time to market is long compared to information technology. In the decade 2006–2015, cleantech companies raised just $960 million (not counting Better Place), or just 4% of the total.[33] Some of the biggest companies in the field have been sold off to foreign companies, including Solel, which was acquired by Siemens in 2009 and later shuttered, and Luz Industries, which went bankrupt in 1991 and was later reincarnated as an American company, BrightSource Energy. One of the biggest failures in money terms for Israeli high tech was a cleantech start-up: Better Place, which raised some $850 million to build a global electric vehicle refueling network. Although it received the technology backing of Renault and investment by a host of blue-chip companies, it ran through nearly all its cash before its business gained any traction. In Israel, which due to its small size and isolation from surrounding countries was an ideal prototype market, Better Place sold only a few hundred cars and experienced serious delays in getting its network of battery-recharging stations in operation.[34] Better Place's collapse was also indicative of the problems Israeli start-ups face in managing the transition from small, R&D-focused business into bigger organizations and the different work culture they require.

Despite the success and recognition that Israel's start-up industry has achieved over the past two decades, there are indications that it has reached a plateau. Trends in the country's information and communications technology (ICT) sector are better barometer of the sector's performance because they exclude sectors such as pharmaceuticals that are included in most definitions of high tech, and they show that ICT growth has lagged the overall economy in recent years. From 2011 through 2015, ICT as a percentage of business-sector gross domestic product was essentially flat, ranging between 10.4% and 11.5%.[35] ICT exports as a percentage of total exports of goods and services grew sharply to 19.1% of the total from 14.7%, but exports of ICT products were flat all that time. The increase was entirely in the export of R&D services, namely multinational R&D centers based in Israel and start-ups. In any event, much of Israel's ICT manufacturing exports are generated by Intel's semiconductors plant in Kiryat Gat, whose output rises and falls based on the product cycle for chips. Its sales don't reflect the strengths or weaknesses of the Israeli tech sector. Looking at the high-tech industry from the perspective of venture capital invested in any single year as a percent of GDP, it has varied between as little as 0.5% in 2009 and 2010 to a more typical level 0.8–1.1% in most years.[36] While the number of companies formed has nearly doubled in recent years, employment in the tech sector has not been rising. And while tech company fund-raising showed big growth in 2014–2015, the increase represented mainly big growth in late-stage deals by more mature companies. Investing in seed and early-stage start-ups as a percentage of total investment has been on a decline over the 2006–2015 decade.[37] The slowdown in the growth of the start-up business model may well reflect a situation where the start-up industry has not only reached the outer limits of what the economy can produce in terms of engineers but also in terms of entrepreneurs and managers. More certainly, access to venture capital funding has become more difficult as the structure of the VC industry has changed—with fewer domestic funds that were traditionally readier to invest in seed-stage start-ups and the concomitant growth in the presence of foreign venture funds. Israel is not alone in seeing its tech sector plateauing: The USA, the industry for more than a decade following the tech bubble of the early 2000s showed virtually no growth in terms of VC investing and deals.[38] A surge in 2014–2015 was already over by 2016. Since the USA doesn't face the same constraints of entrepreneurial and engineering talent as Israel does, largely because it can rely on immigrants, this suggests that the start-up phenomenon globally may be reaching its natural limit relative to the rest of the economy.

Israel counts very few mature technology companies that have sustainable, evolving businesses, with large workforces engaged in the whole gamut of corporate activity from logistics and finance to marketing and human resources management. Among them are Amdocs (25,000 employees, $3.6 billion in sales in 2015), Check Point Software Technologies (3900, $1.6 billion), Nice Systems (3300, $927 million), Orbotech (2300, $753 million), Stratasys, a US-Israeli company (2500 employees, $695 million), TowerJazz (4600, $961 million), and Mellanox (1920, $658 million), but nearly of all these companies were formed in the 1990s or earlier and predate the rise of the Israeli tech sector in its current form.[39] In the last decade, only a handful of companies have emerged as long-term or potentially long-term players opting to go public instead of putting themselves up for sale, among them Wix (a maker of web-development tools formed in 2006) and CyberArk (an information security company formed in 1999). An especially likely candidate to remain independent, Mobileye, which was formed in 1999, went public in 2014 and had emerged as a major Israeli player in auto-tech, sold itself to Intel in 2017. But as the numbers show even the biggest of Israel's technology companies remain relatively small players in terms of the broader economy and in terms of generating employment. Of the 10 biggest employers in Israel's technology sector, three are defense companies (albeit companies that specialize in military electronics) employing a combined 36,000, and three are local units of foreign multinationals, with 14,000 on their payroll. The four top Israeli non-defense tech companies together employed just 11,300.[40] The only significant manufacturers of high-technology products in Israel are foreign companies, and even those are few, given Israel's high labor costs and low productivity.

Where Israel has to some extent exceeded is in the kind of manufacturing where a premium is put on flexible thinking and problem solving and less so on discipline, hierarchy, and other values that characterize industrial plants built around old economy production lines.[41] This is most evident with Intel's success at its semiconductor fabrication plant in Kiryat Gat. Another area where it has been successful is industrial equipment, where companies such as KLA-Tencor, HP Indigo and Orbotech (the first two US companies and the latter Israeli) make machinery used in the production of semiconductors, LEDs, printed circuit boards, and other manufacturing processes and advanced digital printers. Ed Mlavasky, who is one of the pioneers of the Israeli venture capital industry, says Israel could succeed in manufacturing products where costs are less of an issue, citing as an example Xjet Solar, which develops ink-jet printers for use in the manufacturing of solar photovoltaic cells. "If you look at the price per pound of a electronics, if it is consumer electronics—how much does it cost? Not very much. But if you take a $1 million or $2 million machine that does optical inspection of silicon wafers, it weighs tons."[42]

On the whole, however, Israel is not competitive in manufacturing. A study by the consulting firm Applied Economics for the Office of the Chief Scientist (OCS) in Israel's Industry, Trade, and Employment Ministry found that Israeli productivity in technology industries has fallen relative to Finland and especially the USA over a 13-year period to 2008.[43] By 2000, the peak of the global technology bubble, the level of risen to 190% for all three countries of their 1995 levels. All three subsequently experienced a drop, but Israel's was sharpest of them all, and it never recovered. By 2008, productivity per worker had grown to 450% its 1995 level in the USA and 385% in Finland. In Israel, the rate ranged between 160% and 165% from 2001 on. At about 80% the US level, Israeli technology industry productivity lagged the USA less than overall industrial productivity, which is about 55% the US level.[44] A productivity gap like that certainly acts as an obstacle on Israel's efforts to go beyond the research and development focus of start-ups.

The problem facing Israeli tech companies aspiring to create large and sustainable business can be summed up by the relative performance of Israeli start-ups in comparison with older, more mature tech companies. As a survey by the OCS notes, the two sectors are so different from one another that they have to be measured by different variables.[45] Mature companies were gauged by the usual indicators, like output, exports, employment, stock market performance and valuations connected with mergers and acquisitions, and other financial transactions. By comparison, start-ups were measured by venture capital or similar fund-raising, exits, the number of new start-ups and the value of start-up fund-raisings and fund-raising by venture capital funds themselves. What the index constructed by OCS showed was, in effect, two different industries. Mature tech companies showed a steady decline in the 2006–2009 periods. From then on, they recovered but as of 2013 the rebound never exceeded much beyond the levels of 2005. Start-up companies suffered a more severe decline than their mature peers in 2006–2010, but they rapidly recovered in 2011 and staged a second rebound in 2014.[46] While the global financial crisis and the resulting deep recession certainly was a factor in the overall decline in 2008–2009 of Israeli high tech, which is so heavily geared to overseas markets especially the hard-hit US and European economies, the decline predated the crisis. For more mature tech companies, the gradual emergence of the global economy from recession had little impact, which suggests that their problems were not linked to the business cycle as it was to the Israeli business environment and a business culture that is detrimental to the development of big businesses that can compete globally. Start-ups ultimately benefitted from the crisis: The downturn in global R&D spending in the crisis years left multinational companies few options but to acquire technology quickly to make up for lost time by buying start-ups for their IP.

The failure of Israel's tech sector to create larger companies may be feeding back into start-up sector by deterring investors from putting money into promising young companies. Nahumi says the failure of the Israeli high-tech industry to create large companies is a factor that has deterred investment by foreign venture capital funds at a time when they are accounting for an increasingly larger share of technology investment. "If I make an investment in a company that can be sold for $50 million, the return I make for the fund is too small and it creates an opportunity cost. The firm is very involved in the company we're investing in it .... It doesn't make sense for $1.2 billion fund to make such effort for a $50 million exit. Therefore, by definition a firm like ours is trying to fund those companies who want to be big, who want to go public, who want to be self-contained ... Most of the foreign investors are shooting for very large companies in Israel. What's been unfortunate, in the last 10 years only one company I know of has made [such] an exit."[47]

Israel's inability to create and sustain mature tech companies comes at a cost to the economy in terms of employment, as figures from the Israel Association of Electronics and Software Industries show.[48] Companies with 20 or fewer employees accounted just 1.5% of total employment in the Israeli high-tech industry. Those employing between 20 and 50 people for 3.1%, those between 50 and 100 for 6.7% and companies with 100–200 each on their payroll for 14.6%. Nearly three-quarters of all employment in the industry was from companies with 300 or more employees. It is not just simply an issue that smaller enterprises create few jobs but that they create a very narrow range of jobs. While some smaller enterprises are engaged in real business, with productions, logistics, and sales, in the technology sector they are more typically start-ups wholly or principally dedicated to research and development and employ a very limited range of professions, namely for engineers and a small cadre of support personnel. Larger companies by their very nature are true businesses, not only employing more people but also people with a wider range of skills and educational levels. They also employ a wider range of age groups because they require a workforce with managerial and other business experience while start-ups are focused younger people with the latest engineering skills. Since the prospects of these start-ups evolving into bigger companies are extremely poor the wider economy won't eventually benefit with the creation of more and more varied jobs from the investment in IP.

The Israeli technology industry's failure to evolve beyond an intense focus on R&D is borne out by a survey of outcomes for venture capital-backed start-ups in Israel, the USA and Europe taken for Israel Advanced Technology Industries trade association that covers the years of the global high-technology bubble, the slump that followed and the modest recovery that followed until the global financial crisis set it in 2008. Among companies formed from 1996 through 2006, Israel had a slightly higher 44.4% rate of companies remaining independent than the 41.5% in the USA.[49] However, the rate of failure for Israeli start-ups was a much higher 33% rate than the 23.8% for Americans. That most likely reflects in part the more difficult business environment in Israel where start-ups have to rapidly develop a global market, with the attendant risks and added costs, because the domestic market is virtually nonexistent. But the high failure rate also is symptomatic of a "sell it or close it" ethos among Israeli start-up entrepreneurs. "If you are designing your company in order for it to be sold, you end up with a company that has technology but doesn't have sales, marketing or production," explains Zohar Zisapel, one of Israel's veteran technology entrepreneurs and the controlling shareholder of the RAD Group of companies. "If you're successful you're okay, but it's a bit a gamble and when you come to sell it and there are no buyers, then you are stuck. If you are building a real company ... there's always the alternative of keeping on going."[50]

Going public is for all intents and purposes the only route for a high-tech company intent on remaining independent and growing. This is particularly the case for Israeli companies where an overseas listing enhances its profile among customers and potential partners and awards it a foreign, or more usefully, an American imprimatur. This in part explains the reluctance of Israel's best companies to list on the Tel Aviv Stock Exchange. But 25 years after the Israel tech industry came into its own, there is no discernible trend of Israeli start-ups pursuing IPO path. In the 15 years after 2000—the final year of the global high-technology bubble when 38 companies raised more than $1.5 billion in the USA—the number of Israeli IPOs rarely exceeds the single digits in any one year and in four of the years it was nil.[51] There have been brief surges of IPO activity, for instance in 2007 and 2014, but that reflected favorable market conditions that made the mergers and acquisitions route less attractive rather than a trend toward staying independent as the industry matures. Israeli high tech's presence on Wall Street has actually declined: Between 2000 and 2013, some 66 Israeli companies delisted from the Nasdaq Stock Market, which is the chief venue for Israeli high-tech companies trading in the USA, compared with 59 that were newly listed.[52] To be fair, Israeli technology companies are not alone in having avoided the IPO route, a trend that has been various ascribed to regulatory changes, most notably the 2002 Sarbanes-Oxley Act,[53] and to changing business environment, where small companies can no longer generate competitive high levels of profitability.[54] More recently, the 2012 Jumpstart Our Business Startups Act (JOBS) in the USA contains provisions that exempt emerging growth companies (those with less than $1 billion total annual gross revenues the year before their IPO) from key Sarbanes-Oxley requirements. In fact, there was a notable increase in Israeli IPOs in 2013–2014, but by the following year, the trend had run its course.

Regarding mergers and acquisitions, the IATI data found that American technology companies are more likely to be absorbed in an M&A sale (29.2%) than their peers in Israel (17.3%) or Europe (18.9%), but there is a critical difference in the outcome of these transactions. The typical American company is likely to be sold to another American company—either through a merger of roughly equal companies or an acquisition merger into a much larger entity. Either way, the mergers and acquisitions process acts to create larger companies and helps sustain the large enterprises that already exist. In Israel's case that process of aggregation and strengthening of industry players has not occurred. A database of Israeli high-tech M&A activity for the six years 2005 through 2010, covering a total of 235 transactions, showed that 78% of the acquisitions involved the sale of Israeli company to a foreign buyer.[55] In fact, that figure understates the extent of foreign acquisitions and how frequently Israeli start-ups become absorbed into overseas enterprises. Among those M&A transactions where information was available on the price, foreign buyers dominated the M&A process even more decisively. Of 43 sales where the price was in excess of $40 million, only five were by an Israeli buyer. In fact, for the great majority of the 235 sales, the acquired company was valued at less than $40 million and in many cases in the single-digit millions, meaning they were bought while they were still in the research and development stage or had not advanced very far in developing marketing and sales. In other words, the companies were being acquired for their intellectual property and/or for their research and development teams, not because of their business performance.

Many, if not most, Israeli start-ups acquired in cross-border M&A deals become local R&D centers for the company that acquired them. As a result, foreign research and development centers play a major role in the Israeli Silicon Valley. Numbering about 240 and counting among them the world's leading technology multinationals,[56] foreign companies are a major source of employment, account for most of its impressively high levels of business research and development spending and similarly make up a large part of its technology trade surplus. Virtually every major company in global high tech has a substantial R&D presence in Israel (although rarely anything else), among them Intel, Microsoft, Apple, Google, Facebook, IBM, Texas Instruments, Siemens, and Motorola. The world semiconductor industry has an unusually large presence with some 150 design centers employing approximately 20,000 people.[57] All told, foreign R&D centers accounted for more than 54% of all business R&D spending in 2012, five times the national average for European Union countries and far in excess of Ireland's 25%, despite Ireland's renown as a center for multinational R&D and other corporate operations.[58] Moreover, in 2012, multinational R&D centers accounted for 27% of all patents registered in Israel and IBM was the company was the single greatest number among all foreign and Israeli firms.[59] The vast foreign R&D presence in Israel is a testament to Israeli innovative prowess, but it provides further evidence of Israel's inability to leverage its intellectual property into businesses that manufacture products or provide services based on it. The high proportion of foreign R&D to total spending also reflects the low levels of industrial R&D in the non-technology sectors of Israel's economy and the small scale of Israeli technology companies. Even if start-ups are dedicated almost entirely to R&D, they typically employ less than 100 people altogether. By comparison, Microsoft employs about 600 people in Israel, Siemens 800, SAP 800, and IBM some 2000.[60] Thus, foreign companies in Israel both contribute in a significant way to the country's R&D prowess, but they embody many of the same employment profiles that start-up companies do. While a few companies have manufacturing and sales operations in Israel, most of the foreign R&D in Israel is conducted by local subsidiaries that engage in no other significant operations. Far from correcting the imbalance in Israeli high tech toward research and development, they exacerbate it.

When the Israeli high-tech industry came into its own in the 1990s, the natural expectation was that among the hundreds of start-ups that were being formed every year a few would grow into big companies, mimicking the process in Silicon Valley. It didn't happen, and some 25 years later, Israel remains a "Start-Up Nation" of tiny companies that haven't been able to fully leverage the country's innovative abilities into employment and exports.

Notes

1. *Start-Up Ecosystem Report 2015*, 24.
2. *IVC 2016 Yearbook* (Tel Aviv, 2016), 24; and *IVC 2011 Yearbook* (Tel Aviv, 2011), 21.
3. *IVC 2016 Yearbook*, 12.
4. OECD, *Science, Technology and Industry Scoreboard 2015* (Paris, October 19, 2015), 97.
5. Central Bureau of Statistics, *Statistical Abstract of Israel, 2016* (Jerusalem, 2016), Table 18.17.
6. Ibid., Table 18.18.
7. Central Bureau of Statistics, "International Aspects of Israeli Civilian Research and Development," 2. Technology BOP encompasses all commercial transactions related to cross-border technology, such as money paid or received for the use of patents, know-how, technical and industrial research and development carried out abroad.
8. Ibid., 3. In Finland, it was 2.9%, in Sweden 2.8%, and Switzerland 2.3%. In much larger economies, it is not surprisingly considerably smaller—0.6% for the USA and 1.5% for Germany.
9. *IVC 2016 Yearbook*, 12 and 27. Although the figures aren't fully comparable, total foreign direct investment into Israel during those years was $92 billion, according to Bank of Israel figures.
10. Source: *Israel High-Tech and Venture Capital Directory* and news reports. The companies acquired in 2008 were:

Diligent $1.9 million	Mediguide $4.7 million
Fraud Sciences $11.3 million	Onaro $2.0 million
Kidaro $3.7 million	Qumranet $3.1 million

Employment data were derived from news reports and government agencies.

11. Source: Tel Aviv Stock Exchange and company financial reports based on average market capitalization for the year and year-end payroll reports. The companies surveyed are:

Bezeq $309,000	Makhteshim Agan $1.03 million
Elbit Systems $207,000	Osem Investments $286,000
Israel Chemicals $1.7 million	Paz Oil $617,000
Oil Refineries Ltd. $967,000	Teva Pharmaceuticals $962,000

It should be noted that the petroleum and chemicals sectors enjoyed particularly strong profits in 2008 because of higher agricultural-commodities prices so that their market valuations were commensurately high as well.
12. *IVC 2016 Yearbook*, 17, and author's calculations. Life sciences accounted for 22% of all investment over the decade, and software and Internet 21% each.
13. Ibid., 12, PwC MoneyTree Historical Trend Data and author's calculations.
14. Ibid., 12.
15. Ibid., 15–16.
16. One example is the emerging market for civilian drones, where Israel was a technology pioneer in military applications. However, the efforts of Israeli start-ups to enter the civilian sector have been undermined by export restrictions and a strict licensing reimage. See Eliran Rubin, "Israel's Drone Industry Angry over Draconian Defense Export Restrictions," *Ha'aretz*, November 30, 2016.
17. Amitai Ziv, "Cyber Power: Sales of Israeli Companies 10% of All World Transactions" (Hebrew), *TheMarker*, June 25, 2015.
18. Amitai Ziv, "Virtual Iron Dome: Israeli Cyber Will Save Digital Lives" (Hebrew), *TheMarker*, March 24, 2015.
19. John Reed, "Israeli Tech Start-Ups Find Open Lane in New Automotive World," *Financial Times,* September 13, 2016; and Eliran Rubin, "As Vehicles Go High-Tech, Israel's in the Driver's Seat," *Ha'aretz*, October 7, 2016.
20. Emanuel Timor and Liran Hason, "Mapping Israel Start-Ups—Automotive 2.0," Vertex Ventures, June 6, 2016.
21. *IVC 2011 Yearbook*, 16, and author's calculations.
22. Interview with Dror Nuhumi, partner, Norwest Venture Partners, January 23, 2012.
23. Oren Freund, "Babylon Speaks the Tongue of the Markets," *Ha'aretz*, June 7, 2012.
24. Inbal Orpaz and Orr Hirschauge, "Game over in Download Valley?" *Ha'aretz*, August 23, 2013, quotes an industry executive: "Perhaps for a year or two revenues will continue growing, but it's not a genuine

industry.... Companies like Google and Facebook are starting to fight back. In response, Download Valley companies are looking for ways to make their existing assets liquid and create new assets. What's happening now is an attempt by the companies to maintain the revenues derived from the existing models and at the same time reinforce and increase other activity – until they can forgo the original business."

25. Amir Teig, "The Ad Tech Employment Engine: More Than 16,000 Workers in 600 Companies" (Hebrew), *TheMarker*, August 24, 2015.
26. *IVC 2016 Yearbook*, 17 and author's calculations.
27. Israel Advanced Technology Industries, *Israel's Life Science Industries 2015* (Tel Aviv, May 10, 2015), 17.
28. *IVC 2016 Yearbook*, 17 and author's calculations.
29. Interview with Jonathan Goldstein, September 10, 2013.
30. *The PwC Israel 2014 Hi-Tech Exit Report*, undated.
31. Israel Advanced Technology Industries, *Israel's Life Science Industries* (Herzilya Pituach, Israel, May 10, 2015), 40.
32. Israel Advanced Technology Industries, *Israel's Life Science Industries*, 37.
33. *IVC 2016 Yearbook*, 17, and author's calculations.
34. Daniel Schmil, "Better Place Asks Court to Appoint Liquidator as Last of Its Cash Runs Out," *Ha'aretz*, May 26, 2013.
35. Central Bureau of Statistics, Output, Gross Domestic Product and Exports of Information and Communications Technologies Industries, undated.
36. *IVC 2016 Yearbook*, 17, and author's calculations.
37. Ibid., 15. It should be noted that figures on VC investment may overstate the extent of the decline as new sources of start-up capital have grown in recent years in Israel, particularly angel investors and crowdfunding.
38. PwC-National Venture Capital Association, MoneyTree Report, Historical Trend Data.
39. Company reports.
40. Tzahi Hoffman, "All the Numbers Behind the 261,000 High Tech Workers" (Hebrew), *Globes*, July 17, 2011. The biggest defense employers are: Israel Aerospace Industries, Elbit Systems, and Rafael. The biggest foreign employers are: Intel, Hewlett-Packard, and IBM.
41. David Rosenberg, *Cloning Silicon Valley* (London, Pearson Education), 107.
42. Interview with Ed Mlavsky, chairman emeritus, Gemini Israel Ventures, January 23, 2012.
43. Ora Coren, "Research: Productivity in Industrial High Tech in Israel Declined in the Last Decade Relative to the United States" (Hebrew), *TheMarker*, July 8, 2012. The study used 1995 as a baseline with productivity per worker set at 100%.

44. Ibid.
45. Office of the Chief Scientist, *Innovation in Israel* (Jerusalem, April 13, 2015), 3.
46. Ibid., 8. Mature companies on the OCD index scored an 0.25 in the first year of 2005 and were at 0.50 in the final survey year of 2013. Start-up companies started at 0.50 in 2005 and scored 1.74 in 2014.
47. Interview with Dror Nuhumi, partner, Norwest Venture Partners, January 23, 2012.
48. Ora Coren, "The Billion Dollar Question: What's an Exit Compared with Eternity" (Hebrew), *TheMarker*, August 7, 2012.
49. EGP Applied Economics, *The Venture Capital Industry in Israel: An Empirical Analysis of Economic Potential* (Tel Aviv, May 2009), 8. The survey examines the status of companies at the end of 2008 broken down by the year of their founding. The figures cited here aggregate data for 1996–2006 for the USA and Israel and 1999–2006 for Europe. Data for 2007–2008 were not included since too little time had passed to make the data relevant (i.e., for companies, formed in 2007–2008, more than 97% were "independent"). A separate survey, the *Israeli Startup Success Report 1999–2014* conducted by Israel Venture Capital Research with more up-to-date figures covering 1999–2014, found the average failure rate in any given year averaged 46.8%.
50. Zohar Zisapel, speaking at the Israel Innovation Marathon, March 17, 2012.
51. *IVC 2010, 2011* and *2016* Yearbooks, 29, 43, 44, and 45.
52. Wikipedia, "List of Israeli Companies Formerly Quoted on the Nasdaq."
53. The US Sarbanes-Oxley Act raised the regulatory costs of going public while the Global Settlement 2003 between the US Securities and Exchange Commission and 10 large investment banks restricted the scope of analyst coverage, hence stock market valuations of companies no longer enjoyed the same extent of coverage. Both reforms had the great impact of smaller publicly traded companies.
54. Xiaohui Gao, Jay R. Ritter, and Zhongyan Zhu, "Where Have All the IPOs Gone?" April 3, 2012.
55. Dan Yachin and Oren Raviv, *HTIA Annual Review Israel ITC Industry*, 17–18, 24–25, 30, 34–35, 41–42, 49–51, 58–59. The database contains individual transactions. Estimates for the extent of cross-border M&A and others are the authors. Israeli companies, of course, engage in cross-border acquisitions but the number is much smaller. The IVC Research Center counted 17 such transactions in 2009 valued at $6.08 billion (of which $5.4 billion were by Teva Pharmaceutical Industries) and 35 in 2010 valued at $382 million.

56. Israel Ministry of Industry, Trade, and Employment, *Invest in Israel, Statistical Profile* (Jerusalem, August 2011).
57. Interview with Shlomo Gradman, entrepreneur and chairman, Israeli High-tech CEO Forum, June 26, 2013.
58. OECD, *Science, Technology and Industry Scoreboard 2015*, 109.
59. Inbal Orpaz, "Who's at the Forefront of Israeli Innovation?" *Ha'aretz*, August 8, 2013.
60. *Invest in Israel, Statistical Profile*.

CHAPTER 6

Entrepreneurism

In spite of their country's reputation as a nation of technology start-ups, Israelis are less likely to launch a new business in general than people in most of the developed and emerging market economies surveyed in the Global Entrepreneurship Monitor (GEM). In 2010, only 5% of Israel's adult population was engaged in a newly established enterprise, which put at 47th place among 60 countries surveyed.[1] Although the rate grew in subsequent years to 10% in 2013, that still left Israel at 36th in the rankings among 67 countries.[2] On the one hand, living in a relatively developed and wealthy economy that provides opportunities for salaried employment and career advancement in either established businesses or government, Israelis should be less likely to form businesses than their peers on poorer countries. That means Israel's ranking could be expected to be relatively low, since two-thirds of the countries surveyed are lower-income economies. On the other hand, at Israel's current stage of economic development, its rate of entrepreneurship should be growing as it makes the transition to an innovation-driven economy, where continuously changing technology and consumer habits, as well as easier access to finance and market knowledge, create incentives and opportunities for people to start-up their own businesses. That, however, has not been the case in Israel.

Although the rate has fluctuated in GEM surveys taken over the previous decade, on the whole the rate of new-business formation has been in decline since 2002, when 7.1% of the population reported being engaged in some type of new business, and only showed a sudden sharp

D. Rosenberg, *Israel's Technology Economy*, Middle East in Focus,
https://doi.org/10.1007/978-3-319-76654-6_6

rise in 2013. Among the 24 countries defined as innovation-driven by the GEM survey for 2012, Israel's 6.5% rate of early-stage entrepreneurial activity placed it at just 11th,[3] albeit an improvement 13th place among 22 two years earlier.[4] But the relative improvement in the rate of new-company formation is not impressive taken in the context of larger economic developments: Israel's economy was enjoying relatively strong growth in the two survey years, which should have been a strong incentive to start a new business, while most of the innovative economies as defined by GEM were only slowly climbing out of the deep recession of 2008–2009.

A survey by the Organization for Economic Cooperation and Development (OECD) confirms Israel's relatively low rate for generating new enterprises. Using 2007 figures, before the onset of the global recession brought down the rate for new-enterprise creation across the developed world, Israel had an employee enterprise birth rate (the proportion of new enterprises to total enterprises) of 9.1%.[5] By that measure ranked Israel 17th among 21 countries surveyed. Israel's ranking should be taken with some caution. First, not all countries report new-business formation the same way and so country-by-country comparisons can be misleading. Second, the highest rates of new-born enterprises tend to be among the OECD members with the lowest incomes and least-developed economies, confirming what the GEM report says. Nevertheless, even among high-income countries, Israel still ranked a relatively low eight among 12. Moreover, its rate of new-born enterprises declined in the next four years to 8.5% in 2011, at a time when Israel's economy was showing strong rates of economic growth.

Israel's low rate of entrepreneurship—and certainly its decline in recent years—can't be ascribed to its early history of socialism and state-directed business. Even during the years of the British Mandate and into early statehood, a large parallel economy of private business always existed. Moreover, Jews have historically engaged in small-scale trade and commerce so the great majority of Israelis trace their origins to a culture of business that remains strongly evident in diaspora communities. The most likely explanation for the low rate at which Israelis start new businesses is that government regulations and bureaucracy impose excessively high costs on business by demanding lengthy and time-consuming procedures and rules that frequently change and are often unpredictable. The World Bank's Doing Business report, which assesses the business environment in 190 global economies with an emphasis on the needs of small and medium-sized enterprises, confirms that: It ranked Israel 52nd, putting it below all the major developed economies of Western Europe, North America, and Asia except Greece (61st).[6] Another factor is the excessive concentration of many business sectors in the Israeli economy, as was discussed in Chapter 4, although that arguably has more impact on aspiring small businesses seeking to expand than the creation of wholly new enterprises. The most serious problem is bureaucracy, where the World Bank found that Israel does particularly badly in categories such as issuing construction permits, getting electricity, registering property, and tax issues. An OECD survey of the competitive environment among member countries confirms this, ranking Israel second worst among 33 countries survey for product market regulation and for barrier to entrepreneurship.[7] To the extent that Israel has made great strides in opening its economy over the last two decades—for instance, opening the telecommunications, financial services, and aviation industries to more competition—its reforms have been more beneficial to big companies than to small business, with the exception of start-ups that have been largely left alone. The OECD survey found that the administrative burden on sole-proprietor firms in Israel was about 20% heavier than on the average for OECD countries while for start-ups and corporations the burden in Israel was close to the average.[8] Israeli business suffers not just from an excess of regulation but the poor way regulations are constructed and inefficient enforcement. The rise of populist economics in the wake of the 2011 social justice protests exacerbated the problem. Not only were policymakers under increased pressure to respond to demands from the media and elected officials for draconian regulation, reforms that would help new small businesses were largely ignored because they were big or important enough to attract public attention or public pressure. The only exception to the rule of excessive burdens on small enterprises is the tax rate, as against tax policy, which doesn't appear not to be a significant factor in entrepreneurship: At 25% in 2012, Israeli statutory corporate tax income rates were about average among the countries surveyed and they have since fallen.

Even when the rate of new-enterprise formation is narrowed to businesses employing medium or high technology, Israel is laggard. In 2007, according to the GEM survey, Israel was in second place among 42 surveyed countries, with 16.5% of Israeli entrepreneurs reporting they were engaged in medium- or high-technology enterprises.[9] That rate fell precipitously in subsequent years to as low as 2.9% in the global recession year of 2009 and recovered in 2010 to 5.1%, a rate that put Israel in

34th place among 67 countries surveyed. It fell again in 2012 to 4.5%, pushing its rank down to 37.[10] How could an economy with a justified reputation for generating new start-up companies rank so poorly in the GEM rating? The GEM study doesn't define medium- or high-technology enterprise, but presumably the category encompasses not only the classic start-up company (one developing an innovative new product or service based on proprietary research and development) but any new business that makes use of a product, service, or business concept new to the market it is serving. The first category of entrepreneur can build his or her firm on the foundation of Israel's abundant engineering talent and investment capital available for start-ups. He or she has no reason to be concerned about barriers to entry in the local market because it is too small to factor into the nascent start-up's business plan. Indeed, the lack of correlation noted in the previous chapter between the availability of venture capital or broader economic conditions and the rate of new start-ups in Israel suggests that start-up entrepreneurs do not even trouble to examine business opportunity as ordinary entrepreneurs weighing the launch of a new business. They regard their prospects for succeeding as so highly speculative that they are disconnected from the conventional business considerations, especially from domestic market environment. But the second sort of entrepreneur, one who is considering whether to introduce an innovative new product and service to the domestic market, has to contend with the inefficiencies of the domestic market, whether it is an absence of true competition, an excess of regulations, or under-developed infrastructure. One sector where this dichotomy is evidenced is in fintech, where a handful of Israeli start-ups have introduced innovative insurance offerings in the USA. The Israeli start-up Lemonade, which raised $60 million in the first two years after it was formed in 2015, offers low-cost home insurance by employing artificial intelligence, insights from behavioral science and a bot to communicate with clients. It began selling policies in New York State in 2016 and was seeking licenses in most of the rest of the USA in 2017. The company operates a research and development center in Israel but doesn't sell policies there and never sought to use its home country as a test market. Other Israeli insurance tech start-ups like Hippo and Next Insurance have also pursued the US market rather than starting to Israel and there are competitions for insurtech start-ups to encourage other entrepreneurs to enter the segment. But the Israeli domestic market is difficult to penetrate, dominated by five companies that in the segment for house insurance controlled 67% of the market in 2015.[11]

Unlike the conventional nascent enterprise and entrepreneur, the great majority of Israeli start-ups and the people who make decisions for it—its founders, managers, investors, or employees—do not have any expectations of building a sustainable business out of their enterprise, far less any aspirations of creating a large enterprise that will lead or play a significant role in a major market. Rather than being directed toward the ordinary goals of a new business—developing a commercially viable product or service, producing it, building sales and marketing operations, management structures, ensuring the product or service remains contender in the market by continuous product development, and generating profits—the Israeli start-up is typically dedicated to a research and development program geared to a specific innovation and market need. It often entails no long-term business strategy, much less creating an organization that can support it. A profile of Israel's high-technology labor force by the Ethosia Human Resources for the financial daily *Globes* provides evidence of how heavily weighted the industry is to R&D and how little employment it provides in other kinds of jobs normally found in the business sector. The survey excluded many job categories the government includes in the technology workforce to make it more tightly focused on jobs ordinarily regarded as high tech.[12] Of the 169,000 jobs the survey could categorize, close to a third were in software and hardware engineering and design. Among other categories, the industry employed more physicists (4433) than it did people in sales (4158) or finance (3702). More were working as experts on algorithms (3262) than in sales support (2833), as production managers (2572) or in business development (2399) by way of a few comparisons.[13] "The high technology industry focuses its resources mainly on research, development and sales, while the remaining jobs are providing support for these central functions .... The figures indicate a clear trend – Israel is a country of architecture and planning while manufacturing and even development are done in lower-cost countries," Eyal Solomon, Ethosia's managing director, told Globes.[14]

That R&D focus is shared by many, if not most, tech entrepreneurs around the world. But, eventually the most successful start-ups evolve out of it, certainly if they see themselves as businesses that will keep growing and developing. However, that doesn't seem to be the case with Israeli start-up entrepreneurs. Compared to their peers in Silicon Valley, Israeli entrepreneurs begin with more modest goals and end with more modest outcomes. As noted earlier, the Start-Up Ecosystem Report found that Tel Aviv ranked fifth among 20 global technology clusters in its Start-Up Ecosystem Ranking.[15] While Israel ranked third globally for talent, for company performance the report ranked Tel Aviv sixth and for "market reach" 13th, calling Israeli entrepreneurs less "ambitious" in its 2012 report. They were 46% more likely to address smaller markets of less than $1 billion than their Silicon Valley counterparts and, not surprisingly, their average rate of growth and level of revenue is lower.[16] Israeli entrepreneurs are also on average quicker to sell their companies than at least their European peers. A survey for the third quarter of 2014 by Dow Jones Venture Source found that the average time to exit for Israeli start-up was 3.95 years. German start-ups were nearly as hasty at 3.97 years, but UK companies waited an average of 6.41 years, French companies 6.66 years, and Swedish companies 9.03 years.[17] Moreover, the time to exit for Israeli companies has dropped sharply from an average of 8.59 years in 2009 and 5.5 years in 2013.[18] Neither the Start-Up Ecosystem nor the Venture Source reports address the question of why Israeli entrepreneurs set for themselves less "ambitious" targets and exit earlier than others. One answer could be that because they are based in a small country distant from major markets and shadowed by constant political and security risks, Israeli entrepreneurs face a much more difficult business environment than their American, European, or Asian counterparts that constrains their ambitions. But a bigger factor is almost certainly that the typical Israeli entrepreneur is building his company in anticipation of an exit and that requires focusing on a single, salable innovation that can be developed quickly and relatively inexpensively. The strategy meshes well with the needs of multinational companies buying start-up companies in Israel and elsewhere in the world, which far more often than not are seeking intellectual property first and products second. They have little interest per se in acquiring a full-fledged business with customers, branding, and market share. Multinationals already have the skills and organization to develop a business from a start-up's IP—and more likely than not they will integrate it to an existing product suite—and don't need its other business functions.

Gilad Tuffias, co-founder of the Tel Aviv-based technology incubator TechLoft that served as a home for early-stage Internet and mobile companies, describes the process that his start-ups typically undergo in a sharp trajectory from idea to exit. The majority of companies are initially financed by families and friends, which enables them to develop their idea for the first several months of the business.[19] Then, they might seek to raise somewhere in the neighborhood of $100,000, the amounts that TechLoft itself will provide, along with temporary office space and services like accounting. The budding entrepreneurs work for another couple of months, and if they are successful will embark another on another round of financing for between $500,000 and $1 million. "The main thing is the short cycles in businesses in Internet and mobile sector in terms of generating traction and going through all the stages from idea to alpha, beta and getting it out into the market and ultimately turning it into a successful business or making an exit." The goal of his portfolio companies is an exit somewhere between 18 months and three years, he says. In fact, a survey of some 10,000 start-ups formed in Israel between 1999 and 2014 found that 17% of those deemed "successful" (just 480 of the total) had been started without any external financing at all,[20] i.e., they never had to raise any institutional capital that has required them to develop the managerial and financial structures that a venture fund or corporate investor would have required.

Accelerators like TechLoft represent the sharp edge of the Israeli start-up phenomenon. Accelerators (and to a lesser extent technology incubators, their close cousins, which offer support services but a less intensive networking environment) reduce the cost and risk of starting up a new technology company and provide a shared environment for seed-stage entrepreneurs, each "enterprise" consisting of one or two desks in a large shared space with similar early-stage companies, many if not most of them first-timers. The sharing culture, which is encouraged by a large kitchen/lounge area and frequent networking events, leads to collaborations and exchanges of knowledge. Google has dedicated a floor at its research and development center in Tel Aviv to an incubator for as many as 20 pre-seed enterprises with places for up to 80 budding entrepreneurs hosted for several months before being set off on their own. Microsoft's accelerator program in Israel, one of eight the US company operates around the world, has hosted 70 Israeli start-ups since it was launched in 2012, with 80% of them raising a combined $90 million of venture capital. Even the Tel Aviv municipality has become part of the phenomenon, turning an underused library branch into a facility for early-stage entrepreneurs. Twenty-eight people who get approved by a screening committee have access to "hot" desks, a conference room, a panoramic view of the Mediterranean Sea, and a standing invitation to events and meet-ups. All told there are at least 50 accelerators and incubators, in Israel with sponsors that include multinational corporations, venture capital funds, universities, and nonprofit organizations, a few local businesses and as free-standing businesses.[21] For big technology companies, the accelerators give them an eye into up-and-coming technologies emerging from the start-up world that could threaten their businesses if they don't fail to latch on to them quickly. Multinational companies see them as a critical enough investment that accelerators have been developed across Europe as well as in the USA. From the point of view of the Israeli economy, however, by creating a powerful link between the very youngest start-ups and big multinational companies, accelerators serve to reinforce the early-exit phenomenon in Israel.

The question is why Israel has been so successful on the one hand at generating start-up companies and on the other hand has failed to develop large, sustainable enterprises even as the industry marks its third decade. The answer lies in a constellation of Israeli cultural mores and values, which easily adopted the start-up culture first developed in California's Silicon Valley. The purely cultural factors that drive the start-up phenomenon and deter the development of bigger enterprises in Israel are discussed in Chapter 8; in the meantime, the discussion here is limited to the institutional and micro-economic factors that constrain Israel's start-up industry.

Israel adopted the start-up culture early and rapidly in the 1990s, only a decade or so after California's Silicon Valley pioneered the concept. It has held fast to it ever since. Annalee Saxenian describes the start-up-centered environment in Silicon Valley that enabled it to emerge as the unchallenged center of the global high-technology industry, even though Boston's Route 128 at the outset easily rivaled it as America's other big technology cluster.[22] Both centers hosted large numbers of big technology-oriented companies, were home to big concentrations of engineering and scientific talent, and benefited from a wealth of universities and research institutes. But, in contrast to Route 128 firms, which were vertically integrated, self-contained, and secretive, Silicon Valley's tech companies were collaborative. They developed partnerships and shared technology among themselves, as well as with investors and academics. Silicon Valley companies were open to employees setting out of their own by starting up new companies and often provided them with financial and technology assistance. This relaxed attitude toward institutional structures extended to inside the enterprise, where relations between employees were less hierarchical, information was shared widely, and respect for established practices was minimal. In other words, the high-tech industry as it developed in Silicon Valley was as much about business culture as it was about technology. Many of Saxenian's insights have become so widely acknowledged that they have become clichés, but they are worth repeating only because the Silicon Valley ethos is one and the same as Israel's. In fact, it is hard to find other countries, including those that enjoy the same foundations of a knowledge economy, such as a highly educated workforce and good universities, that have succeeded to the extent Israel has in adopting the Silicon Valley ethos for their own. In Israel, the growth and development of the high-tech sector didn't quite parallel what happened in Silicon Valley because in Israel there were few big companies to engage in collaboration or mentor young start-ups. However, there was a tradition of collaboration in the defense sector, usually under the aegis of the government. Moreover, as a small country with a population of engineers who served in the army and studied in the small number of university engineering programs, an informal environment of sharing and collaboration already existed.[23]

Thus, Israel generates far more start-up companies on a per capita basis than any other country in the world, including Europe and the USA. Using Europe as a benchmark, *The Wall Street Journal* using data from its sister company Dow Jones Venture Source, Israeli start-ups had raised 15 times more capital on a per capita basis than their European counterparts from 2003 through third quarter of 2013.[24] Israeli companies raised nearly two-thirds more venture capital per capita than American companies and 3–4 times as much as Swedish and British start-ups. On a measure of venture capital-financed investments, Israel also came out far ahead of other advanced economies: It enjoyed nearly seven times the number of transactions per capita than Europe and about a third more than the USA and Sweden. Not only is start-up activity more intense in Israel, Israel's technology cluster is a purely domestic phenomenon, which testifies to the extent it is a homegrown phenomenon. This distinguishes Israel's technology start-up sector from Silicon Valley and the technology clusters emerging in Europe, where foreign entrepreneurs are a major and welcome presence. It is true that non-Israelis play a critical role as investors with the majority of venture capital coming from overseas. Moreover, inside start-up enterprises themselves, foreigners often head up marketing and sales functions that require a familiarity and expertise in overseas markets that Israelis have difficulty matching. But the core activities of an Israeli start-up—the engineering and entrepreneurial aspects of the company—are conducted almost wholly by Israelis.

Indeed, the Israeli start-up industry is so thoroughly intertwined with Israeli character and culture that it is difficult for non-Israelis to penetrate it. Even though the country faces a critical shortage of skilled personnel, the number of non-Israeli entrepreneurs in Israel is tiny as is the number of engineers or others in senior positions within start-ups.[25] Part of this is due to severe restrictions on people entering the country on work visas (Israel also has a highly specialized "expert worker" visa program, but industry executives say it is very difficult for applicants to win approval) and to Israel's uncertain security environment.[26] But another critical element is the difficulty of non-Israelis fitting into an industry that is so deeply embedded in Israeli culture, despite its firmly global orientation. Ed Mlavsky, a veteran venture capitalist, notes that of the 25 Israeli leading Israeli technology companies all have Israeli CEOs, despite their being global businesses. "Bringing in foreigners to run companies has not worked very well for Israel in the past," he says. "In the early days of VC industry it was absolute article of faith that when company reached the stage they had products to sell and the target market was almost always an America, you needed an American to run the American operation and the American operation would be most of the company. But that turned out to be absolutely disastrous."[27]

The reverse side of the coin is that Israelis fit easily into Silicon Valley, New York City, Boston, and other places where the start-up culture thrives in the USA. An estimated 50,000–100,000 Israelis alone live in the area centered on Palo Alto, California, the epicenter of Silicon Valley.[28] Many of them work for American tech companies but at least 110 local start-up companies have been founded by Israelis.[29] In Massachusetts, Israeli-founded businesses raised nearly $700 million in venture capital in 2010–2012 in 73 transactions, representing over 6% of all venture capital funding in the state.[30] In both states, Israelis represent a disproportionate share of immigrant start-up entrepreneurs—in California, they accounted for 3% of all start-ups formed by immigrant entrepreneurs in 2006–2012 and in Massachusetts 16%.[31] Nationwide in America, Israelis were the sixth-largest group of immigrant entrepreneurs, accounting for 3.5% of the total.[32] In certain high-tech sectors their share is unusually disproportionate, comprising 7% of all immigrant founders of computer/communications start-ups and 11% of all new immigrant companies in the semiconductors field.[33] As do expatriates from India and China, two other large groups of foreign nationals present in US high-tech centers, Israelis certainly take advantage of the networking opportunities presented by a large community of fellow nationals. But that can't be the only factor. The interconnected nature of the technology business, as Saxenian demonstrates, would make it impossible for a single ethnic or national group to succeed solely by relying a network of compatriots. Rather, the outsized Israeli presence is evidence that Israelis easily fit into the start-up culture, whose features are similar wherever it exists. Thus, the culture and education that have given Israelis the ability to start-up companies in Tel Aviv serve them just as well in Palo Alto or Boston.

The venture capital funds that provide the majority of capital to the industry are often blamed for the phenomenon of early exits. The funds have limited life spans and need to create liquidity and show a return to their investors. In recent years, as it has become more difficult for VCs to raise capital, the pressure to generate returns quickly has mounted as has the preference to conserve capital by shedding by selling companies in their portfolio that will likely require further funding. Fund managers, however, deny they are the principal factor behind companies to sell. "When a VC fund is getting close to the end of the fund, it will try to do some fundraising. But the VCs don't run the companies – this is a big illusion," says Mlavsky, whose Gemini Israel Ventures is one of Israel's oldest VC funds.[34] "There is a lot of pressure for the entrepreneurs and the other stakeholders, and there is pressure from the VC. But the pressure from the VCs is uneven because they need to show good results only when they are in fundraising mode. It's not the only time, but it's more acute."

Mlavsky and others suggest that the prime driver for early exits in Israel is the entrepreneurs themselves who make a multifaceted cost-benefit calculation that encompasses not just the prospective financial return from selling their company early but the personal cost going forward involved in building a company for the medium and/or long term. "Being entrepreneur is a great way to earn of living, not to live a life. The typical entrepreneur works ridiculous hours and every day of the week. They don't see their families grow up. When somebody waves a fat check at them, it's very tempting when they do the arithmetic. They are really set for life," Mlavsky says. Looking at the personal calculations an entrepreneur might make, he explains: "'If I wait another year, I may be able to get three times [what I am being offered now], but can I wait another year?'" He recalls a situation where a Gemini portfolio company in the enterprise software sector received an offer to be bought for under $100 million. In that case, the venture capital fund was proactive in discouraging the sale; a stand Mlavsky admits is unusual. "The founders and principals were ready to take it, but what we did – which is something many VCs don't do – was to enable the principals, specifically the CEO, who was also the founder, to cash in on some of his stock on the next round of financing so he was able to show something for his effort."

Michael Eisenberg, of the Silicon Valley-based venture capital fund Benchmark Capital, suggests another psychological dimension. "This is a country where people want to be their own boss. This is a hypothesis, not a statement of fact: It may come from the fact that people take a lot of instruction in the army, so when you get out you want to be your own boss. A two- or four-person start-up that doesn't scale is comfortable both from a lifestyle perspective and a career-satisfaction perspective. Then there is economic perspective. There is less of an appreciation of the impact of the size of the overall pie versus my piece of the pie. Silicon Valley entrepreneurs understand that the size of the overall pie is the primary driver of wealth creation. 'I need to own 20% of this company in order to create wealth, versus owning 1% of a Facebook.'"[35] Eisenberg says that if fund managers tended in the past to pressure their portfolio companies into early exits, they are less inclined to do so now. "For the first decade of the VC industry we had a lot of local funds who were uncertain they could raise the next round [of capital]. They wanted to put runs on the board, which means selling out faster to get the exits. Also, they didn't have the experience to take things the whole way," he says. "In the last four to five years, with the entrance of foreign funds, they are bringing more experience and patience on the ground to play for the big outcome."

In fact, there is evidence of Israeli entrepreneurs' growing willingness to let their start-ups grow into sustainable businesses, although as of 2016 the change is not overwhelming. A survey of the high-tech industry by the business information firm Dun & Bradstreet found that the number of "big" tech companies, defined as employing more than 100 people, had grown in the five years to 2016 to 385 from 298.[36] Looking at M&A deals, in 2014–2016, in number terms about 30% involved one Israeli company buying another rather than a foreign company acquiring an Israeli one.[37] However, the growth in the number of big companies by D&B amounted to an average annual increase of 4% and big tech companies by the D&B standard comprised just 6% of the 6650 and were dwarfed by the 4750 start-ups. Moreover, the value of initial public offerings—as clear a signal as any from a company that it plans to remain independent—over the four years through 2016 was tiny compared to the number of M&A exits. In some years IPOs amounted to just a few tens of millions of dollars, according to IVC data.

One proximate cause for the failure of Israeli start-ups to evolve into sustainable enterprises is a shortage of appropriate engineering talent, the area conventionally regarded as Israel's forte. To some extent, this is a matter of pure size: Israel's entire engineering workforce numbers in the tens of thousands, which provides an absolute barrier to how many any single company can employ locally. But that barrier is not as high as it might appear: "We have a lot of engineers – don't misunderstand me – but these engineers exist in a few technology areas," says Eisenberg.[38] "There are integration labs, big integrators like [the Israeli firms] Matrix and Ness, but these companies aren't suited for going to start-ups and scaling them. Then there are tons of people who have built products on the Microsoft stack in the government and the military. For the most part those are in enterprise software deployment. They are not used to scaling the Internet business at the pace they need to." The Israel Defense Forces, which in effect serves as the country's biggest training program for engineers, have long relied on the Microsoft platform and that, in turn, has become the standard in government and for much of industry, even though in many sectors elsewhere around the world open-source software in the norm. More recently, the IDF has begun using the open-source software used by start-ups. For example, its developers now use PostgreSQL, MongoDB, and Neo4j as an alternative to Oracle and since 2015 training for software developers includes programming languages like Python, as well as others for Internet and mobile development like Java, HTML, and Android programming.[39] However, it will take time for these changes to percolate through the private sector.

Meanwhile, however, the problem persists. Eisenberg quotes Avishai Avrahami, chief executive officer of Wix, a start-up that enables users to build flash Web sites: "My business grows at over 10% a month but my competitors will catch up with me because I cannot find Java and Ruby on Rails talent in Israel. I cannot find any engineers with experience in scaling big data centers for many customers. I am willing to pay top wages but I cannot find the talent. I am flying to Russia to find or buy talent."[40] As of 2014, Wix has two development centers in Eastern Europe, in Vilnius and Dnipropetrovsk. But outsourcing to countries such as India, the Ukraine, and even the Palestinian Authority is by no means a solution. It undermines one of the Israeli industry's greatest strengths—the ability to form small and stable teams of people working closely toward a common goal—while at the same time, it demands the kind of sophisticated management skills, organizing and supervising large numbers of people over distant geographies and different cultures, where Israelis have faced the greatest challenges.

Another obstacle to building sustainable enterprises seems to lie in the Israeli focus on technology and engineering at the expense of business execution. Israeli start-ups, like their peers elsewhere around the world, rely on some form of technology leadership to drive their business, which typically gives them a time advantage of two to four years before their competitors before the gap, the technology becomes commoditized or the standard is overtaken by something new.[41] But they are less likely than their American counterparts to successfully develop a strategy of evolving product development, which would entail skill sets involving market intelligence, strategic management, and accessing a deeper and wider range of engineering talents. Instead, they change product niches, rather than try to compete, or put themselves up for sale, ideally while their technology remains valuable. "If you look at the past, Israeli companies did very well when tech barrier to entry was really high, as opposed to a very effective business execution," says Dror Nahumi, a general partner at Norwest Venture Partners, a US venture capital manager that has been active in Israel for many years.[42] "If you take an American company and an Israeli company building the same thing, rarely will Israeli company have better business execution than the American company, its being far away from the market and having a management team more product-oriented than sales- and marketing-oriented." This has been the case for sectors such as telecommunications equipment, semiconductors, and very complex software for enterprise, but the phenomenon has grown more intense with Internet and mobile. "Products that start-up companies create these days are easy to duplicate, so it comes down to business execution, which makes it much more difficult for Israeli companies to compete. It doesn't mean you can't find those that will, but it more difficult. It's not natural for Israeli companies. They are far from the market and didn't create sales and marketing DNA to do that."

Israel suffers from being so geographically distant from trend-setting markets, traditionally the USA, despite the strong cultural connections between the two countries. But, especially with the Internet and the heavy traffic of Israelis between their country and Silicon Valley, it is difficult to explain the Israeli technology industry's failure on this account. In fact, says Nahumi, the marketing barrier for Israeli companies is a function of bigger cultural challenges they face, the first in assigning much value to communication and image and the second in establishing and managing large organizations. Those shortcomings are a function of Israeli society, which puts a premium on communicating directly without much subtlety or strategy. For start-up companies, this arguably serves as a useful cultural asset because within the business itself ideas are exchanged and thrashed out freely, quickly and directly among a small staff all of whom know each other personally and, broadly speaking, share the same values and attitudes. But the flipside of this cultural phenomenon is a devaluation of the skills involved in the more nuanced communications entailed in addressing mass markets, especially as it involves marketing and communicating across different cultures. The prevalent attitude among Israeli start-ups, despite two decades of experience in multiple cross-border markets, is that a product will sell itself because of its superior technology rather than because it addresses market needs and/or is positioned and branded properly. This helps explain why Israeli companies traditionally succeed more often in developing and marketing products designed for business applications rather for consumer markets. Business markets are not only smaller and more focused but are more likely to be driven by fundamental technology than by branding.

In fact, one major and potentially revolutionary change has occurred in recent years in terms of the marketing barrier for Israeli companies, namely the emerging field of Web analytics. Web analysis collects, measures, and analyzes traffic from social media, e-commerce, search engines, and a host of other channels to gain insights into Internet usage on a mass scale. This has created a new realm of global marketing based primarily on technology rather than on interpersonal communications skills or cultural knowledge. As a result, some of Israel's most successful start-ups of the past five years, including companies like Wix (Web site building), Waze (a navigation app sold to Google in 2014), and Gett (a ride-sharing app), have broken the barrier Israeli start-ups have traditionally faced with big consumer markets, each of them variously addressing markets for end-users, self-employed professionals and freelancers, and small businesses. Amir Yarkoni, CEO of an online marketing company Seperia, suggests in a blog on Israelis comparative advantage in Web analytics that the emerging field makes use of Israeli strengths in "entrepreneurship, analytics, daring, and collaborative culture."[43]

A second obstacle relates to the management of large organizations. In the typical evolution, a seed-stage start-up is managed and operated by a small team of people who know each other personally, work together closely and informally. They are directed toward a single goal of developing a particular technology, often led by a CEO whose talents are motivating people by force of personality or by an inspiring vision. As the company evolves, so does the range of tasks it assumes, as do its goals—and its workforce grows accordingly. Investors require financial and other reporting; institutional structures are formed; the scope and timeline of development work expand; and logistics, production, marketing, and financial functions become an integral part of the operation. The company has multiple and competing goals involving many more people. That is typically when the Israeli start-up begins to stumble: The business is no longer driven by technology innovation, rather the innovation itself becomes subject to organization and procedures, which requires administrative and management skills. Everywhere in the start-up universe, Israeli included, only a few start-up entrepreneurs successfully make the transition. In Israel, however, the problem is exacerbated by the paucity of large technology companies. The pool of chief executives and other senior managers from which growing start-ups could potentially draw talent and experience from is virtually nonexistent. "If you look at the greatest 20 companies Israel has created and have been acquired, a lot of these companies had very little revenue," says Nahumi. "They were acquired for the technology, not the business. The outcome is that there is a shortage of managerial teams that know how to build great business, not just great technologies .... You need management who knows sales and marketing and building great business. But we don't have those any more. Those who did build big companies before, they're not available to build new companies. Not all of them are ready to go and fight in the trenches by running a new company."[44]

The kind of entrepreneurism that thrives in Israel is unique to the high-tech business. It starts on the assumption that ordinary business considerations don't factor in starting the business to begin with and that the entrepreneur and the company need not acquire the skill sets of manufacturing, marketing, finance, and human resources that a growing, developing business will need. Unlike an ordinary young business, start-ups have a ready way to monetize their businesses by being acquired—an option that most small businesses don't have—because their primary, if not exclusive, asset is their intellectual property. The start-up enterprises that might aspire to maturing into bigger, sustainable businesses face the obstacles of the small size of the Israeli labor pool for engineering and other professions, a tendency to focus on technology over branding, and a paucity of managers with experience leading large enterprises. Cultural factors, to be disucssed in Chapter 8, factor in a very big way as well. The end result is a thriving culture of start-ups with neither the incentives nor the inclination to create big, sustainable enterprises.

Notes

1. Ehud Menipaz et al., *GEM 2010 National Entrepreneurship Report* (Beer Sheva, Israel, 2010), 17. The figures refer to total early-stage entrepreneurial activity (TEA) as defined by GEM. A new enterprise is defined as a business in one of two stages—started up in the previous 12 months or one in business for between 4 and 42 months and has been paying salaries. The survey in Israel was based on responses from 2007 people aged 18 to 65 representative of the population.
2. Ehud Menipaz et al., *GEM Israel 2013 National Summary* (Beer Sheva, Israel, undated), 9. The 2013 survey included 2039 respondents in the same age group.
3. Siri Roland Xavier et al., *Global Entrepreneurship Monitor, 2012 Global Report* (March 27, 2015), 26.
4. Menipaz et al., *GEM Israel 2013*, 19.
5. Organization for Economic Cooperation and Development, *Entrepreneurship at a Glance 2013* (Paris, August 5, 2015), 5. Employee enterprise businesses are one that employs at least one other person as against people who are simply self-employed.
6. World Bank and International Finance Corp., *Doing Business 2012* (Washington, DC, 2012), 6.
7. OECD, *The 2013 Update of the OECD's Database on Product Market Regulation* (Paris, March 31, 2015), 16 and 18. On a scale of 0–6 (0 being least restrictive and 6 most restrictive), Israel scored 2.15 versus 1.49 on average for OECD countries for product market regulation and 2.50 for barrier to entrepreneurship.
8. Ibid. Data appear in worksheets appended to the main report.
9. Menipaz et al., *GEM 2010 National Entrepreneurship Report*, 30–31.
10. Menipaz et al., *GEM Israel 2012 National Summary*, 2. In any event, the numbers involving high-tech entrepreneurship are so small relative to the total population and the total entrepreneur population—in Israel's case, technology entrepreneurs made up 5.1% of the total entrepreneurship rate of 5.0% in 2010—they say little about entrepreneurialism generally.

11. Finance Ministry, *Capital Markets, Insurance and Savings Division, Report of the Commissioner, 2015* (Hebrew) (Jerusalem, updated), 5.
12. The Central Bureau of Statistics, using OECD measures, includes job categories such as production workers in the pharmaceuticals industry. The Ethosia survey excludes those, but does count R&D workers in the drugs industry.
13. Tzahi Hoffman, "All the Numbers Behind the 261,000 High Tech Workers" (Hebrew), *Globes*, July 17, 2011.
14. Ibid.
15. *Start-Up Ecosystem Report 2015*, 24.
16. *Start-Up Ecosystem Report, 2012*, 21.
17. Amir Mizroch, "In Israel, Startups Race Toward Exit," *The Wall Street Journal*, December 22, 2014. It should be noted that at least vis a vis German start-ups, Israeli companies generated a much higher return to investors, at 6.2 fold, compared with 2.54, in approximately the same timeframe.
18. Ibid.
19. Interview with Gilad Tuffias, co-founder, TechLoft, January 23, 2012.
20. Israel Venture Research, "Israeli Startup Success Report 1999–2014," January 28, 2015.
21. Israel Advanced Technology Industries, online database.
22. Annalee Saxenian, *Regional Advantage: Culture and Competition in Silicon Valley and Route 128* (Cambridge, MA, 1994).
23. Dan Brenitz, *Innovation and State: Political Choice and Strategies for Growth in Israel, Taiwan and Ireland* (New Haven, CT, 2007), 42–43.
24. Ben Rooney, "Ben, How Entrepreneurial Is Europe?" *The Wall Street Journal*, November 11, 2013. The exact figures (Europe = 1.00) are:

	Relative venture-backed Funding	Relative venture-based deals
Israel	15.13	6.69
U.S.	9.61	4.45
Sweden	4.09	4.63
UK	3.16	2.83
Germany	0.92	1.02

The figures do not take into account angel and other funding. The survey encompassed business and finance, information technology, and consumer services start-ups.
25. John Reed, "Israeli Entrepreneurs and Immigration: 'Start-Up Nation' Stifled," *Financial Times*, November 11, 2014.
26. Amir Mizroch and Orr Hirschauge, "Foreign Entrepreneurs Face Israeli Immigration Hurdles," *The Wall Street Journal*, September 23, 2014.
27. Interview with Ed Mlavsky, chairman emeritus, Gemini Israel Venture, January 23, 2012.

28. Inbal Orpaz, "'Israeli Mafia' Thrives in Silicon Valley but Can't Escape High Cost of Living," *Ha'aretz*, January 29, 2014.
29. Companies listed on the Web site of Israeli Executives and Founders Forum as of June 2015. The list only includes IEFF members and therefore undercounts the real number.
30. David Goodtree, *The Massachusetts-Israel Economic Relationship* (Boston, December 2013), 5.
31. Vivek Wadhwa, AnnaLee Saxenian, and F. Daniel Siciliano, *Then and Now: America's New Immigrant Entrepreneurs, Part VII* (Kansas City, MO, October 2012), 22. For comparison's sake, note that Israelis comprise 0.11% of the world population.
32. Ibid., 3.
33. Ibid., 22, 23.
34. Interview with Ed Mlavsky.
35. Interview with Michael Eisenberg, Benchmark Capital, April 15, 2012.
36. Dun & Bradstreet Israel, "The High Tech Industry—The State of Affairs," January 2017.
37. Israel Venture Capital Research, "Israeli High Tech Exits at $10 Billion in 2016: 93 M&A Deals, 8 Buyouts and 3 IPOs," January 3, 2017; "2015 Israeli High Tech Exits Hit $9.02 Billion—Up 16% from 2014 Proceeds," January 11, 2015; and " 2014 Israeli High-Tech Exits Reach $6.94 billion," January 6, 2015.
38. Interview with Michael Eisenberg.
39. Inbal Orpaz, "The IDF's Apps Store" (Hebrew), *TheMarker*, May 18, 2015.
40. Michael Eisenberg, "The Hummus Manifesto—Part 1."
41. Brenitz, 90.
42. Interview with Dror Nuhumi, partner, Norwest Venture Partners, January 23, 2012.
43. Amir Yarkoni, "Why Digital Marketing Is a Critical Growth Engine for Israeli Start-Ups," Seperia (blog), October 28, 2013.
44. Interview with Dror Nahumi.

CHAPTER 7

Human Capital—Tangibles

Unlike traditional industry clusters, which coalesced around the close proximity of natural resources, access to markets and/or on a foundation of older craft traditions, the phenomenon of Israeli high tech is entirely a function of the qualities of its human capital. The industry isn't reliant on natural resources of any kind, Israel is distant from the world's major markers for high tech, and the skills and intellectual abilities the industry requires have little connection with the needs of an twentieth-century industrial economy, much less its antecedents in pre-industrial crafts. Israel's universities have a relatively long history of academic excellence that, together with successive waves of educated immigrants, has given the country a highly educated population by international. Moreover, the armed forces have served as a training ground for technology innovators and entrepreneurs. But Israel's educational achievements on a formal level are not overwhelmingly superior to those of other developing countries—and in some respects, they critically lag behind them—and by themselves can't account for the country's ability to have spawned such a successful high-tech industry. What distinguishes Israel in terms of its human capital is its singular history, which has created a constellation of societal characteristics—a tendency to risk-taking and teamwork, resistance to hierarchy and routine, and a culture of openness and contentious debate—that have made it a seeding ground for the start-up companies that are at the core of its knowledge economy. Thus, it can be said that Israel's human capital asset base has two components. One consists of tangible factors that can be measured in educational and

D. Rosenberg, *Israel's Technology Economy*, Middle East in Focus,
https://doi.org/10.1007/978-3-319-76654-6_7

scientific achievements. The second component, and the one that has made the critical difference between Israel and other developed economies in terms of its ability to generate innovative high-tech companies, is an array of intangibles that comprise the Israeli character and defy the usual measures of an economy's human capital. These are discussed in Chapter 8.

When analyzing the impact of educational quality and achievement, it is important to recognize that the high-technology start-ups that are the core of its knowledge economy are an elite phenomenon. High tech plays an important role in the economy and has become in the minds of policymakers, the public, and the national industry. But start-ups employ relatively few Israelis and don't need to draw on the education, skill, and talents of the great majority of the population. Indeed, as it is structured—a mass of start-up companies employing small numbers of engineers—Israeli high tech doesn't need a large population of skilled labor, certainly not in the wide range of abilities larger businesses would need in management, finance, marketing, logistics, and the like. Other areas of Israel's economy need and draw on those skills, but it is telling that apart from high tech, few other Israeli industries have shown themselves to be globally competitive. That failure extends to areas, such as finance and media, where human capital plays a decisive role in competiveness. In fact, 47.7% of Israel's workforce was employed in knowledge-intensive job in 2013, putting it at No. 7 in the world as a percentage of the total workforce and about at the mid-range among the world's most advanced economies.[1] But the figure probably has less to do with the relative size of Israel's high-tech sector than with the absence of older manufacturing industries or a big natural resources sector. These are major employers in economies like the USA, where knowledge-intensive industries constitute just 38% of the workforce, Australia (44.9%) and Canada (43.7%).

Two other factors also serve to diminish the significance of the preponderance of knowledge-intensive jobs. The first is that there are segments of the economy in Israel that by definition are knowledge-intensive but, as noted in Chapter 4, in practice aren't characterized by the kinds of operations and business strategies normally associated with knowledge-intensive industry. Israel also has a large public sector, which is counted as knowledge-intensive, even though in Israel's case government services are inefficient and by developed-country standards backward. The second factor is that the Israeli workforce has lower level of basic workplace skills than those of most other developed countries. The

OECD's Program of International Assessment of Adult Competencies showed that Israeli workers lagged most of their peers in tests of literacy, numeracy, and problem solving even though Israelis on average have more years of schooling than other countries on average.[2] Moreover, the problem extended across all education levels from those with just a pre-secondary education to those who had completed a master's degree, which suggests that even in knowledge-intensive jobs, Israelis bring inferior skills to the workplace.[3] In other words, the truly knowledge-intensive segment of the economy is concentrated in the much tinier high-tech sector. There, the share of the Israeli workforce employed in the information, communications, and technology (ICT) sector amounted to just 5% of the workforce in 2014.[4] After rapid growth in 2003–2011, the figure has been declining both as a proportion of the labor force and in absolute terms.[5] The decline is mainly due to the sharp contraction of Israel's domestic telecommunications industry after 2012 due to the tariff reforms described in Chapter 4, but even after discounting that segment ICT employment grew in 2011–2014 by just 3.9%.

At the top of the educational achievement ladder, Israelis are extraordinarily well educated by international standards. As of 2012, 46% of the adult population aged 25–64 had a post-secondary degree, nine percentage points above the average for countries belonging to the Organization for Economic Cooperation and Development.[6] Only Canada and Japan enjoyed higher rates. But overall the achievement has been less impressive. Israel has the sixth highest rate of Ph.Ds per capita among OECD countries—5.8 men and 3.9 women per 1000 population.[7] But its rate of awarding doctorates is below the OECD average, at 1.4% of the relevant age group versus 1.6% for the OECD and rates of 3.2 and 2.8%, respectively, for Switzerland and Sweden, the top countries by this measure.[8] Meanwhile, the percentage of Israelis who have completed high school is 85%, which puts it tied for 10th place with Finland and Slovenia among OECD countries.[9] Even that relatively low ranking likely overstates the effective Israel rate of secondary education since it includes a significant percentage of ultra-Orthodox Jews, who spend enough years in school to be counted as achieving an upper secondary education but in fact follow a course of religious study that includes few if any skills appropriate for a modern economy (an issue dealt with in detail in Chapter 11).

Still, even after adjusting for ultra-Orthodox Jews, the average number of years of schooling for the population grew significantly from 10.4

in 1974 to 12.4 in 2011, putting Israel at about in the middle range of OECD countries during those decades.[10] However, Israel's ability to keep pace with the developed world was only enabled by two big waves of highly educated immigrants from the former Soviet Union in the 1970s and 1990s. This is a critical problem for the economy going forward because there aren't likely to be any other major sources of educated immigrants in the foreseeable future. Yet for knowledge-based economies, improving educational levels is a central factor in driving growth. The growth of the educated population has been a major factor behind the Israeli economy's growth over the past two decades, enabling Israel to register big gains in labor productivity and helping to create the human resources foundation for the rise of the high-tech industry. The Bank of Israel estimates that over the last 40 years, the increase in the average number of years of schooling added an average of 0.8% point to annual growth, or 40% of the total, by enhancing productivity.[11]

Apart from the quantitative increase in the number of students enrolled in higher education is the question of the quality of education they are receiving and its relevance to the needs of the economy and high tech in particular. Israel's top universities perform well in global rankings. The Academic Ranking of World Universities conducted annually by China's Shanghai Jiao Tong University ranks six of Israel's seven universities among world's top 500 in 2015 based on alumni and faculty who have Nobel Prizes or Fields Medals, how frequently its researchers are cited, and articles in leading academic journals.[12] Two Israel's universities (The Hebrew University and the Technion) are in the top 100. In sciences, Israel has three in the top 200, in engineering two and in the social sciences three. In the last decade, Israelis have won six Noble Prizes in the sciences, all of them in chemistry, and one Field Medal in mathematics. But it should be noted that these rankings are based on research achievements and reputation, not on the quality of teaching. Nobel Prizes are awarded on the basis of scientific achievements sometimes dating back decades of an elite few and don't say very much about the current overall state of a country's scientific prowess.

There are no international measures of student performance on a post-secondary level, but there is an array of indirect evidence suggesting Israel's position has deteriorated. The number of students enrolled in the country's universities and colleges grew 428% in the years 1973 through 2010.[13] However, enrollments have since plateaued—averaging growth of just 1.5% annually, versus a peak of 8% in the era of great expansion in the 1990s—and are not expected to recover any time this decade.[14] Moreover, during the years of rapid growth in enrolments public spending on higher education did not kept pace, shrinking in inflation-adjusted shekels to 26,500 shekels in 2010 from 82,400 shekels in 1979 (2010 prices) as the state came to rely more on private colleges funded from tuition and other private sources.[15] Meanwhile, the number of senior faculty grew only 40% in those years. The brunt of this decline was felt at Israel's universities, which as a rule accept the best students and conduct most of the country's academic research, especially in the sciences. While the university student population increased by 157% in these years, senior faculty at the universities dropped by half to 62 per 100,000 population.[16] At three of the top universities, senior faculty declined in absolute terms—17% at The Hebrew University, 26% at Tel Aviv University, and 26% at the Technion, Israel's leading engineering school.[17] The student/senior faculty ratio at the universities has grown, for those pursuing a bachelor's degree to 15.6 in 2010 from nine in 1977. At the masters level it increased to eight students from two per faculty member and at the doctoral level to 2.12 from 0.7. The numbers don't just represent more crowded lecture halls and seminar rooms and heavier teaching loads, which inevitably spell less attention to individual students, but also represents a growing reliance on low-paid adjunct faculty who cannot dedicate the same time and resources to teaching or research.[18]

The problems extend further down the education ladder. At 7.3% in 2011 (the last for which there are comparable data), Israel spent a relatively large percentage of gross domestic product on education at all levels—sixth largest among OECD countries—and since then spending has increased considerably.[19] But that is largely due to the fact that Israel has a relatively young population. Measured by spending per student on a purchasing power parity basis, Israel spent what would be expected on students in primary and higher education at its level of per capita GDP (about $7000 and $11,500, respectively), although was considerably less generous in terms of secondary education (less than $6000).[20] Still, given the country's reliance on its human capital resources, this is a disconcerting level of underinvestment, especially given the poor performance of Israeli students measured against their peers abroad. The OECD's Program for International Student Assessment (PISA) exam, which measures the mathematics, reading and science achievements of 15-year-olds in member and other countries, gives Israeli students poor scores in every category, although their performance has improved since the tests were first given in 2000. In mathematics, the mean score for Israeli students was 466, up 25 points from 2006 but well below the OECD average of 494.[21] In reading, their score was 486, 10 points below the OECD average, and in science literacy, it was 470 31 points under the average. In tests of general problem solving, Israelis' mean score was 454, ranking them 34th among 42 countries. Another measure, the Quality of Education Index, which examines a broad range of cognitive skills, rates Israel at a 4.7, below the median of 5.0 and the seventh lowest score among the OECD countries.[22]

The reason for the poor performance is not immediately self-evident. Israeli classrooms have more students than in most other OECD countries, but the teacher–student ratio is close to the OECD average, teaching hours are extensive and early childhood education is widespread.[23] Rather, the problem of Israeli schools seems to be a poor allocation of funds and time. A survey of teachers by the OECD found that Israeli teachers spend less time than most of their peers on teaching (18.3 hours a week, ranking Israel 23rd of 32 surveyed countries) and more of it on discipline and administrative tasks (5.2 hours, eighth among 32 countries).[24] The Bank of Israel report on the low level of skills Israelis bring to their jobs links the problem directly to the poor quality of education, starting with an underinvestment in teaching in early childhood.[25] But, as the bank points out, policymakers to focus on the easier measure of years of schooling rather than on its content.

Irrespective of the quality of teaching, there is a troubling decline in the number of students pursuing science, technology, engineering, and math (STEM) subjects from high school through university. A report from the Office of the Chief Scientist termed it a "funnel" phenomenon of fewer and fewer young Israelis opting to pursue STEM studies as they advance through the educational system starting in high school. In 2009, the number of 18-year-olds in Israel was 118,000, of whom 62,000 took the high school matriculation exam (bagrut) in mathematics.[26] Of those, the number who took five units, the highest level was just 11,000, and the number who excelled, according 85 points, or more, was 6600, or just 5.6% of the total cohort. Those 6600 students in theory are the core of the next-generation technology workforce in Israel, but that number is inevitably reduced going forward as many of these students will opt to study medicine, law, or business. The OCS noted that in 2012, the number of students getting a relevant degree for work in high tech was 4671, compared with a 7000 net new job openings in the tech sector. Thus, the universities are failing to turn out enough graduates and that in turn has led to a perpetual shortage of engineers and others needed for the high-tech industry. Like other advanced economies, this failure partly reflects student preference for subjects like law and business over most STEM subjects.[27] But in Israel, the problem has been exacerbated by funding cuts for institutions of higher education, which has made it more difficult to expand and improve programs and retain faculty. The number of university graduates in fields relevant to high tech declined from 3000 in 2003 to 2100 in 2009, the last year for which figures are publicly available.[28]

Strangely enough, Israel's high-tech sector is the one least affected by the quantitative and qualitative deficiencies of the country's educational system, and there is much evidence that for Israel's start-up sector, the army's technology units serve as an important source of training, supplementing, or even substituting for a formal education, a factor discussed in Chapter 8. The strength of the start-up companies that form the core of Israel's knowledge economy, as well as the multinational research and development centers that are such an important source of employment, is a creative process of innovation. While innovation has to be supplemented by a store of knowledge and skills of the kind acquired through formal education, the weighting of knowledge for start-ups is certainly less than for most other segments of a knowledge economy. In any case, the relatively small size of the sector, and the very limited range of job categories and skills it requires, means it can cherry pick the best and brightest and fulfil its human capital needs. Export-oriented companies in other industries can as well, albeit with somewhat more difficulty given their greater man power needs. For the rest of the economy, the poor state of the schools, the declining quality of graduates, and their skills mismatch have saddled it with a workforce not fully suited to the needs of a knowledge-intensive economy. In fact, the majority of the economy is reliant on low-cost labor often inappropriately trained for their tasks, a function of the rapid increase in the labor force over the last decade.[29] The evidence of this allocation of human resources is evidenced in the dichotomy in Israel industries where Israel enjoys, relative to the OECD average, a large share of people with higher education, such as the export sector, finance and insurance, and technology.[30] Productivity in these sectors is no lower than the average for OECD countries and sometimes exceeds it. By contrast, in sectors where the rate of higher education is similar to other countries, there is a productivity gap to Israel's disadvantage.

In any case, the growth in the number of Israelis with a higher education is stalling, and even going into reverse, as evidenced by the rate of those holding tertiary degrees by age group. In most developed economies, access to higher education has expanded massively over the last decades and younger groups far more have a tertiary degree than their parents or grandparents. Across the OECD, for instance, 39% of the youngest cohort had a degree in 2012, versus 24% of the oldest group aged 55–64.[31] In Israel, by contrast, the rate of tertiary education peaks at 50–51% for those 30–44, but more critically at the far ends of the age spectrum, the rates are the reverse of what is seen in most countries: The youngest cohort of Israelis, aged 25–34, has a tertiary education rate of just 44% while among the two oldest Israeli cohorts, those aged 55–64 and 45–55, the rate is 47 and 45%, respectively. The youngest Israelis' tertiary education rate is just five percentage points higher than the OECD average and only the 10th highest among the 34 member countries. It also is lower than the average for all adults. Part of the unusually low rate can be explained by the fact that many young people delay the start of the higher education until they have completed army service, taken time to travel and/or dedicated a year or more to community service, so that many have not completed their first degree until well into their twenties (the average age of graduates getting their first degree in Israel in 2012 was 29.04, compared with 26.35 for OECD countries on average).[32]

However, more fundamental factors are at work in explaining the Israeli retreat. Like other advanced economies, for now at least, Israel has exhausted the potential for expanding the number of years of schooling. If in the early 1970s the average number of years of schooling was less than high school level, now it is at post-high school level.[33] Moreover, the gap between those in Israel who qualify for university admissions based on high school achievement exams (bagrut) has been closed, so that only if primary and secondary schools succeed in improving student performance is there much room for Israel to increase its tertiary enrolment rate.[34] Two other factors, however, are unique to Israel, namely the contribution of Russian immigration, and the other current—the growth of the Haredi and Arab populations.

The wave of immigration from the former Soviet Union brought some 710,000 people to Israel between 1990 and 1997, a number that was so big relative to Israel's native population at the time that it boosted the working-age population by 15%.[35] The immigrants were unusually well educated, even if their skills were often mismatched to the needs of the economy and it took time for both labor supply and demand to adjust. By one estimate, 60% of the Russians had a tertiary education, compared with just 30–40% of the existing Israeli-Jewish population at the time (and a single-digit figure for non-Jews).[36] The proportion of Israel's population with more than 16 years of schooling jumped from 16% in 1987 to 28% in 2005, although a large part of the increase is also attributable to the expansion of higher education in Israel.[37] Coming from a state-controlled economy and unfamiliar with the ways of capitalism and enterprise, Russians did not become as a rule the founders of start-up companies or the driving force behind the start-up phenomenon that got underway as they were arriving in Israel, but they did make a major contribution to the countries engineering talent—probably in the main by filling jobs in older, established industries, thereby leaving a greater pool of Israelis available for start-ups. In 2007, close to 24% of Israel's ICT workforce were immigrants who had arrived in Israel after 1990, meaning nearly all Russian immigrants.[38] So many immigrant doctors arrived in the 1990s that it raised the number of doctors from 3.1 per 1000 population in a few years to 3.7, one the world's highest levels. In the year 2000, Russian immigrants accounted for 38% of all of Israel's physicians.[39] However, the first generation of Russian immigrants with their unusually high educational profile is gradually exiting the workforce and which is one reason why the percentage of Israelis with a tertiary education is declining.[40]

The second factor contributing to Israel's struggle to ensure a highly educated workforce is the Haredi and Arab minorities, which together comprise about 30% of the population today. The special issues relating to these groups are discussed in Chapters 10 and 11. However, broadly speaking Israeli Arabs receive an inferior education to the general population, which can be seen in lower test scores at the elementary and high school levels and much lower rates of tertiary education. Israeli-Arab schools get less funding than their Israeli-Jewish counterparts, and the fact that they teach in Arabic as against Hebrew puts their graduates as a disadvantage when they go on to pursue a higher education or enter the job market. These are serious problems, which are compounded by discrimination and other factors, but they pale in comparison with the challenge posted by the ultra-Orthodox. While the ultra-Orthodox community puts a great deal of emphasis on education, it is geared almost exclusively toward traditional religious studies. Boys and young men (though less so for girls and young women) get no significant exposure to general studies, including math, science, and English, that would prepare them for productive and well-paying employment a modern economy. Few are equipped to go on to higher education or to army service, which would make up for their shortcomings. A study by the Bank of Israel concluded based on a correlation between education and income that ultra-Orthodox schooling made no contribution to their earnings or job performance abilities beyond someone with a 10th grade education; for Haredi women, their education was the equivalent of having 1.5 years less schooling than their non-Haredi counterparts.[41] Unless, there is a significant change in the ultra-Orthodox attitude toward education in the next four decades and a greater willingness to pursue a general education of non-religious subjects, the Bank of Israel predicts that the increase in the average level of schooling for Israelis will come to a standstill after about 2039, and by 2059 Israel will be in 26th place for average level of education among OECD countries, 10 places below its current level.[42]

The role of religiosity in the economy is not limited to education and its immediate impact on skills training. It also plays a critical role in an individual's and community's worldview in ways that pose a challenge to the core values of a knowledge economy, which requires people prepared to question and challenge assumptions, examine issues and ideas critically, and show little reverence for the past and for tradition. Although traditional Judaism to one degree or another degree incorporates these values, they are less in evidence in the Haredi world, a matter discussed in more depth in Chapter 11. In all events, it is certainly more difficult for these values to freely manifest themselves in a religious environment that values tradition and respect for authority. Yet the rise of Israel's start-up culture has also seen a parallel growth in religiosity, which is evidenced in a 2009 social survey by the Central Bureau of Statistics. It found that 21% of Israelis reported they were more religious now than in the past, a third more than those reporting they were less so. If these trends continue, by 2030 close to half of all Jewish Israelis will be religious by the broad standard that encompasses ultra-Orthodox, Orthodox, and traditional Jews, up from just under a third in 2010.[43] Combined with the problematic trends in formal education, rising religiosity threatens to present a further obstacle to perpetuating the start-up culture at the heart of the Israeli knowledge economy.

The problems with the Israeli educational system and the country's demographics have made themselves felt in terms of labor productivity. Israeli productivity has always lagged behind most of the developed world, and over the last four decades, the gap has widened, in spite of Israel's reputation as a knowledge economy at the cutting edge of global high technology. In 2014, the rate for GDP per hours worked in Israel was just $37.30 (adjusted for purchasing power parity), putting it in the bottom third of the 34 OECD countries and just slightly higher than Greece and Portugal.[44] The Israeli rate was less than half the US rate of $67.40. Over the last 45 years, even as the Israeli economy has made significant strides in deregulation, opening its markets to world trade, eliminating inefficient industrial sectors, and developing high tech as a key industry, the Israeli productivity gap with the leading economies of the Group of Seven has widened. In the mid-1970s, the difference was about $3.40 (in PPP terms using 2005 dollars) but by 2014, it had reached $15.90.[45] Many factors have contributed to Israel's low labor productivity, including the fact that Israeli business under invests in machinery and equipment compared to other developed economies, but another critical factor is the parameters of the labor market. Israel in recent decades has had a low rate of labor force participation, i.e., the percentage of the working-age population holding or actively seeking a job. In recent years, the economy has made enormous strides in raising the figure, but that has come at the expense of higher producivity growth. In 1997, the rate for all Israelis age 15 and older was 53.4%; by 2012, the rate had risen to 63.6%.[46] In fact, measured against economies that like Israel's that include people serving in the army as part of the workforce, the Israeli labor force participation rate was relatively high at 78.7% for those aged 25–64 was well over the OECD average of 76.1%.[47]

The rapid expansion of higher education and the rising education profile of Israelis was a major factor in the growing percentage of workers, most particularly women. But another driver was government policies put into place in the early 2000s making access to government allowances more difficult and restructuring the tax system to put more of the burden on consumption than on income.[48] These policies encouraged people whose levels of education and training are poor into the job market, depressing productivity levels and wages. The evidence of this comes from data showing that after rising in the 1990s, the labor force participation rate among Israelis with 16 years or more of schooling has been flat and has even shown some decline in recent years (as has the rate for those with 13–15 years of education).[49] At the bottom end, those with just 0–10 years of schooling declined sharply in the 1990s, but stabilized in the early 2000s while for those with 11–12 years of schooling, the rate has been rising throughout this period. The bottom line is that the growth in a labor force has been in jobs that require less education, thereby weighing on overall productivity growth. Where Israel has seen the greatest growth in educated workers, and therefore its most productive workers, is among women. Indeed, the rates of education for employed women are higher than for men. But women earn less than their male peers with equivalent credentials, and the gap actually widens as the level of education rises.[50] The reasons for this are subject to debate, but at least some of them are imposed on women, e.g., they sacrifice career to family needs for lack of affordable daycare options and/or they encounter a glass ceiling at work that deters their pay and career-advancement prospects. Whatever the reason, the gender gap represents a large loss to the economy in terms of a large and very productive segment of the labor market.

Given Israel's political insecurity and its lower standard of living, the number of Israelis who choose to leave the country for better opportunities elsewhere is surprisingly small. The level of emigration from the country has declined over the last 20 years, and overall the percentage of émigrés from Israel is not high by developed-country standards. However, the Israelis who due choose to make their home overseas are on average more highly educated and possess greater training and skills than the population at large. Israel is not losing a critical mass of intellectual capital—indeed, the evidence suggests that the movement of Israelis abroad is less a matter of brain drain than "brain circulation," i.e., Israelis leaving the country for extended periods to further their education or gain work or entrepreneurial experience before returning home.

Any discussion of emigration and brain drain is subject to a lot of uncertainty over fundamental data. It is the nature of emigration that people who opt to leave their country of origin do not necessarily inform the authorities. Many people living abroad for extended periods of time do not plan to stay permanently and many who leave temporarily for a short-term job assignment, education or even a holiday end up making their homes abroad. In Israel's case, the large number of immigrants among the population—most notably the wave of more than one million people who arrived from the former Soviet Union in the 1990s—means

Israel has large numbers of people with dual citizenship as well as strong family, cultural, and other ties to other countries that increase the likelihood they will return to their former homelands for extended periods, if not permanently.

Various estimates have been made regarding the number of Israelis living abroad. Drawing data from a variety of sources and taking a few educated guesses, Yinon Cohen, estimated there were 544,000 Israelis living abroad as of 2006.[51] Of those, however, 300,000 were foreign born, many of them Russians who immigrated to Israel because it was the easiest first destination available to them and had few Zionist inclinations, and they quickly moved on to a third country. Immigrants who return to their home countries are not unusual—the rate for those coming to Britain in the decade to 2006 was 40% and in the USA the return rate is estimated at between 25% and 40%. In Israel, the Central Bureau of Statistics says that just 10% of the Russians who arrived in the 1990s wave of immigration subsequently emigrated. A more recent estimate by Israel's Central Bureau of Statistics is close to Cohen's estimate, somewhere between 552,000 and 588,000, not counting children of Israelis born overseas.[52] Taking figures from a survey by the OECD and adjusting them for the fact that Israel has such a large immigrant population, Cohen concluded that Israel had an emigration rate of just 5.85% of its population, not much higher than the 4.9% OECD median and lower than 11 OECD countries, including Ireland, New Zealand, Switzerland, Austria, Finland, Greece, and the UK.[53]

Yinon's figures for the percentage of Israelis living abroad come from the year 2000, but the fact is there's no reason to assume the emigration rate has grown since the rate of emigration has declined since then in absolute terms and even more so relative to Israel's population. Government figures show that since the latest peak 2001–2002 at about 27,000 people per year, the number of emigrants had fallen to 16,200 in 2013.[54] Meanwhile, the number of Israelis returning ranged between 7800 and 11,000 per annum in those years. Added to that, Israel enjoys a relatively large rate of immigration, which means Israel's migration balance is consistently positive. While the rate of immigration is down far from its peak in the 1990s, since the early 2000s it has ranged between 13,700 and 24,100 annually.[55] Israel has no policy favoring immigrants with skills and education—the great majority come for ideological reasons, namely Jews choosing to live in a Jewish state—yet 46% of those who arrived in 2014 had 16 or more years of schooling.[56]

Given Israel's self-proclaimed mission to be the homeland of the world's Jews, emigration has been traditionally looked down upon as a betrayal of the Zionist enterprise. People who immigrate to Israel are known as *olim* (those who are ascending) while those who leave are *yordim* (those who are descending). Attitudes have relaxed over the last decades, and the phenomenon is now often couched more in terms of what is so wrong with Israeli society that it causes so many of its citizens to leave, rather than in blaming those who do. The debate, which is often tinged by political views, variously ascribes the phenomenon to Israel's chronic security problems and the absence of a peace process and/or to narrowing opportunities for the middle class. In all events, it is hard to detect a pattern in emigration rates that would correlate with any of these long-term political or economic factors over the last three decades. Emigration rose in the first years of the Oslo peace process and rapid economic growth, but that increase came in parallel with the wave of immigration from the former Soviet Union and probably reflected large numbers of recent immigrants who moved on to third countries or back to Russia. Emigration figures rose again in the first two years of the Second Intifada and the effective end of Oslo, but they declined long before the violence subsided. In any case, the figure was at least as likely due to the sharp recession Israel experienced in those years as it was to the upsurge in violence. Although the last decade has been a period of growing economic distress for the middle class (a phenomenon discussed in Chapter 9), the rate of Israelis leaving the country for a year or more actually declined. In all events, an emigrant from Israel is more likely to be someone who had earlier immigrated to the country and is therefore less likely to remain given the difficulties of adjusting to a new and different society and the ease of returning to one's home country. Yinon estimates that 300,000 of the 544,000 Israelis living abroad in 2006 were foreign born.[57]

If the overall rate of emigration isn't high, where it most frequently occurs is among the most highly educated, in spite of the rise of the Israeli high technology industry and the business and job opportunities it is has created. In particular, Dan Ben-David, an economist at Tel Aviv University, has documented what he called "an academic exodus unparalleled in scope"[58] attributable to the stagnant growth in senior faculty at Israeli institutes of higher education over the decades. The percentage of Israeli academics working abroad is 10 times or more the rate for any other developed economy, with close to 25% of them overseas in 2003–2004.[59] In 2007, Israeli scholars comprised 0.1% of total senior faculty at American colleges and universities, but in an indication of the quality of intellectual capital Israel was losing in the years prior to that, Israelis accounted for between 1.5% and 1.6% of senior facility in the USA's top 40 departments of philosophy, physics, and chemistry in 2007.[60] In economics, they comprised 3.1% and computer science 3.8%. Israeli physicists teaching at US universities equaled one-tenth the number teaching in Israeli institutions.[61] This brain drain not only deprives the next generation of Israelis of some of the country's best teaching talent but also deprives the economy of academic research with the potential to be commercialized.

On the other hand, the Israeli high-tech sector does not bleed human capital, at least not on a scale that endangers the industry. The country has succeeded in retaining most of its best and brightest, despite the obvious difficulties of working and doing business in Israel from the standpoint of personal security and the lower standard of living. There are no reliable figures on the number of Israelis in Silicon Valley—the most likely destination for an ambitious tech person—although the number often cited is between 50,000 and 100,000, and some say less plausibly 200,000.[62] However, even the low estimate is a large number, considering that total Israeli ICT employment (not counting domestic telecommunications operators) was 153,000 in 2014.[63] Moreover, Israelis are resident in other tech sectors across the US, especially New York and Boston, not to mention in Europe, so the total of lost talent is much larger. Anecdotal evidence of a large Israeli presence is strong, with handful of large-tech companies founded by Israelis, as well as scores of smaller companies and start-ups.[64] Expat Israelis serve in key positions in US technology leaders like Intel, Apple, and Google as well as investors and entrepreneurs. A survey estimated that Israelis accounted for 3.5% of all immigrant start-up entrepreneurs in America, the sixth largest foreign cohort and well out of proportion to Israel's share of the world population or the overall US immigrant population.[65]

For all that, however, many of these Israelis are spending time in the US to develop skills and a network of business contacts that they will eventually use back home either by returning to live there or to start up companies or open research and development arms for their US businesses. Many others are working for Israeli start-ups that have located their headquarters and/or marketing operations in the USA to be close to customers while keeping their core research and development operations back in Israel. On the balance, while Israel's tech sector has certainly lost talent to the USA and to a far lesser extent Europe, the damage is not as great as the raw numbers would suggest.

Short of fundamental changes in education spending and policies, and a successful effort at improving access to education and job opportunities, Israel faces a severe human resources crunch over the next decades. The knowledge economy, especially in Israel, with its focus on innovation and change rather than accumulated skills, is by its nature a youthful one. Yet, one major trend identified by a government report released in 2012 is the inevitable aging of Israeli society. Israel's population is youthful compared to its developed-country peers due to a high birthrate,[66] but the proportion of Israelis over the age of 65 is expected to climb from 9.8% in 2009 to 12.1% in 2019 and 13.8% in 2034 as Israel's baby boom generation of the 1950s and 1960s reaches pensionable age (a figure that is being inflated by the Russian immigration, where the average age of arrivals was about 40).[67] By 2059, the rate could reach as high as 21%. For the broader economy, this raises serious questions about the ability of a working age population shrinking in relative terms to support a growing proportion of pensioners. The answer that policymakers in Israel and other developed economies have proffered is to raise the minimum age of retirement to qualify for government benefits. This would certainly have the effect of reducing the dependency ratio, but it is a less than ideal solution given that a fundamental feature of the knowledge economy is its perpetually changing knowledge and skill requirements, to which older workers inevitably have trouble adopting. With Israel's knowledge economy's focus on start-ups, raising the retirement age is particularly problematic because the industry is at the cutting edge of extraordinarily dynamic business and social trends. Among Israeli start-ups today, and even among larger and older techno logy companies, only 3% of all employees are over age 50.[68]

An equally serious problem is the growth of Israel's Haredi and Israeli-Arab populations, which is discussed further in Chapters 10 and 11. The government's mid-range scenario forecasts that the Haredi population to grow by five-and-a-half fold between 2009 and 2059 to 4.15 million people, increasing its share of the total population to 26% from 10%.[69] The Israeli-Arab share will grow to 3.61 million, or 23.1% of the population from 20.4% as its population grows by 135%. By comparison, Israel's population of non-Haredi Jews and others will grow by less than 50% in those years and their share of the population will fall from 69.7% in 2009 to 66% in 2019 and to just above 50% in 2059. There are serious difficulties measuring the Haredi community, the biggest variable in Israel's demographic outlook because the community has no official designation and encompasses a wide range of religious practices and attitudes. People move into and out of the Haredi community during their lifetimes, further complicating forecasts. But the labor force participation rate among Haredim is low and their education incompatible with the needs of a modern economy. The same issues apply, albeit for different reasons, to Israeli Arabs. The inevitable consequence is that unless the characteristics of the two minority populations fundamentally change—a process that will involve not only investing more in education but also altering deeply embedded social attitudes—Israel's knowledge sector will be deprived of a large and growing pool of human capital.

Notes

1. Silja Baller, *World Economic Forum Global Information Technology Report 2016* (Geneva, 2016), 261. The WEF uses International Labor Organization figures, which define the knowledge-intensive workforce as managers, professionals and technicians.
2. Bank of Israel, "Broader View of Selected Issues," in *Fiscal Survey and Selected Research Analyses* (Jerusalem, October 2016), 16.
3. Ibid., 17.
4. Central Bureau of Statistics, "ICT Sector Estimate for 2014—4% Increase in GDP" (Hebrew), August 12, 2015.
5. Central Bureau of Statistics, "ICT Sector Estimate for 2011—1.5% Growth in GDP" (Hebrew), August 9, 2012.
6. Organization for Economic Cooperation and Development, *Education at a Glance 2014: OECD Indicators* (Paris, October 2014), 44.
7. Organization for Economic Cooperation and Development, *Science and Technology Scoreboard, 2013* (Paris, October 23, 2013), 96.
8. Ibid., 94.
9. OECD, *Education at a Glance 2014*, 43.
10. Bank of Israel, "The Development of Education in Israel and Its Contribution to Long-Term Growth," *Recent Economic Developments, No. 136*, April–September 2013, 24.
11. Ibid., 23.
12. Academic Ranking of World Universities, 2015, Shanghai Jiao Tong University.
13. Dan Ben-David, "Update on the State of Israel's Universities and Its Researchers," in *State of the Nation Report: Society, Economy and Policy in Israel, 2013* (Jerusalem, November 2013), 305.
14. Israel Council of Higher Education, Selected Data for the Opening of the Academic Year—2016–2017 (Hebrew), 2.

15. Ben-David, 305.
16. Ben-David, "Brained Drained," 3, with more recent figures from "Update on the State of Israel's Universities and Its Researchers," 303.
17. Ben-David, "Update on the State of Israel's Universities and Its Researchers," 304.
18. Ben-David, "Brained Drained," 14–15.
19. Bank of Israel, "The Connection Between the Quality of Education and Growth: Israel Compared to the World," in *Recent Economic Developments, No. 139*, October 2014–March 2015, 8.
20. Ibid., 10.
21. Organization for Economic Cooperation and Development, Education GPS, Student Performance (PISA 2012).
22. Bank of Israel, "The Connection Between the Quality of Education and Growth: Israel Compared to the World," 11.
23. OECD, *Education at a Glance 2014*, 444, 446 and 428, respectively. Class size in Israel was the fourth highest among OECD countries at 26.95 on average and actually increased from 26.71 in 2000. The student–teacher ratio was 15.2 in 2012, close to the OECD average. The number of classroom hours for nine years of primary and early secondary education was 8752, versus an OECD average of 7475.
24. OECD, Education GPS, Israel, Teachers and Teaching Conditions, 2013.
25. Bank of Israel, "Broader View of Selected Issues," 29.
26. Office of the Chief Scientist, Economy Ministry, *Innovation Report* (Hebrew) (Jerusalem, April 13, 2015), 22.
27. National Economic Council, *The Skilled Manpower Shortage in High Technology: Recommendations of the Inter-Ministerial Committee* (Hebrew) (Jerusalem, July 2012), 34.
28. Ibid., 36.
29. Bank of Israel, "Labor Productivity in Domestic-Market Oriented and Export Industries: An Analysis from an International Perspective," in *Annual Report 2013*, 65.
30. Ibid., 64.
31. OECD, *Education at a Glance 2014: OECD Indicators* (Paris, October 2014), 44.
32. Ibid., 74.
33. Bank of Israel, "The Development of Education in Israel and Its Contribution to Long-Term Growth," 27.
34. Ibid., 26.
35. Sarit Cohen and Chang-Tai Hsieh, "Macroeconomic and Labor Market Impact of Russian Immigration in Israel," October 2000, 3.
36. Ibid., 32.

37. Amit Friedman and Noam Zussman, "Labor Force Quality in Israel" (Hebrew) (Jerusalem, March 2008), 9.
38. Knesset Research and Information Center (Ilanit Bar), *Main Influences on Immigration from the Commonwealth of States After 1990 on the Israeli Economy* (Hebrew) (Jerusalem, April 24, 2012), 8.
39. Ibid., 10.
40. Among Russian immigrants, the decline in education is particularly pronounced. Among immigrants age 55–64, 51.8% have a post-secondary degree; among the second generation age 25–34 the rate is 30.6%. This is partly due to the late age Israelis typically get their bachelor's degree but also due to the low socioeconomic status of first-generation immigrants relative to their educations credentials. See Liza Rozovsky, "Educated Immigrants from the Soviet Union Boost the Job Marker—Their Children are Under the Average," *Ha'aretz*, December 7, 2015.
41. Bank of Israel, "The Development of Education in Israel and Its Contribution to Long-Term Growth," 21.
42. Ibid., 27.
43. Evengia Bystrov and Arnon Soffer, *Israel: Demography 2012–2013* (Haifa, May 2012), 60.
44. Ben-David, *The Shoresh Handbook on Israel's Society and Economy* (Shoresh Institution for Economic Research, 2015), 3.
45. Ibid., 4.
46. Bank of Israel, *Annual Report 2012*, Statistical Appendix, The Civilian Labor Force and the Participation Rate, 1969–2012.
47. Ibid., 175.
48. Ibid., 179. Compared with most other OECD countries, Israel derives a larger share of its state revenues from value-added and other consumption taxes than it does from income tax.
49. Ibid., 179.
50. Ayal Kimhi, "Trends in the Labor Market: Gaps in Employment Rates and Wages," in *State of the Nation Report 2011–12* (Jerusalem, undated), 142.
51. Yinon Cohen, "Israeli-Born Emigrants: Size, Destinations and Selectivity," *International Journal of Comparative Sociology*, 52(1–2), 2011, 50.
52. Central Bureau of Statistics, "Departures and Returns in 2013 of Israelis Staying Abroad Continuously for a Year or More," August 19, 2013, 6.
53. Cohen, "Israeli-Born Emigrants," 51.
54. Ibid., 5. Exact figures can be found in: http://www.cbs.gov.il/reader/?MIval=%2Fcw_usr_view_SHTML&ID=635.
55. Central Bureau of Statistics, "Immigration to Israel in 2014," May 19, 2015, 2.

56. Ibid., 6. A third had less than 12 years of schooling, but that is because official figures count the number of years an immigrant had upon arriving in Israel, including children.
57. Cohen, "Israeli-Born Emigrants," 50.
58. Ben-David, "Brained Drained," 2.
59. Ibid., 6.
60. Ibid., 9.
61. Ibid., 10.
62. Inbal Orpaz, "'Israeli Mafia' Thrives in Silicon Valley but Can't Escape High Cost of Living," *Ha'aretz*, January 29, 2014.
63. Central Bureau of Statistics, "ICT Sector Estimate for 2014—4% Increase in GDP," August 12, 2015.
64. The Israeli Executives and Founders Forum, a group of Silicon Valley-based Israeli tech professionals, counted 114 Israeli-founded start-up in the region as of October 2015, a figure that likely understates the true number (ieff.us/map/#). A count of Israeli start-ups in New York for the same date shows 262 (www.israelimappedinny.com) and for Boston 89 (www.israelimappedinboston.com).
65. Vivek Wadhwa et al., *Then and Now: America's New Immigrant Entrepreneurs, Part VII* (Kansas City, MO, October 2012), 8.
66. OECD (2015), Fertility rates (indicator and OECD (2015), Young population (indicator). Israel's fertility rate in 2013 was 3.03, by far the highest among OECD countries. In contrast to other OECD countries, Israel's birthrate has risen since 2005. The proportion of its population under age 15 was 28.1%, the second highest behind Mexico.
67. Ari Paltiel et al., *Long-Range Population Projections for Israel: 2009–2059* (Jerusalem, March 21, 2012), 53.
68. Orr Hirschauge, "Profile of Israeli High Tech: A Young Male, Veteran of an IDF Combat or Tech Unit and a University Graduate," (Hebrew), *TheMarker*, March 11, 2013.
69. Paltiel et al., *Long-Range Population Projections for Israel*, 46.

CHAPTER 8

Human Capital—Intangibles

The foundation of a knowledge economy is its human capital, the aggregate of knowledge, skills, abilities, experience, intelligence, reliability, and judgment of a country's people. Usually, these are captured in measures of formal educational achievement, in which Israel excels, although not to a degree that can fully explain the emergence of the thousands of start-up companies that form the core of its knowledge economy. Formal educational achievement can only partly explain the start-up phenomenon in Israel; the rest, and quite probably the great majority of it, is instead attributable to an array of societal characteristics that can't be captured in figures about the percent of the population with tertiary degrees or math and science scores of adolescents in international standardized exams. In Israel's case, these intangibles include a strong distaste for organization, hierarchy, and rules; a propensity for risk-taking; a highly developed culture of teamwork and, as a corollary, a high degree of loyalty to colleagues but also a high degree of candidness and directness in interpersonal relations; and a preference for addressing practical problems and directing energy to defined tasks over theory and organizational discipline. Some of these characteristics, such as risk-taking, play an obvious role in start-up culture and others, such as team loyalty, do not. But to anyone who has observed Israel's start-up culture from up close can easily discern the critical role each of them play.

These national characteristics have arisen from the uniquely Israeli experience. This experience draws from the shared history Israelis have

D. Rosenberg, *Israel's Technology Economy*, Middle East in Focus,
https://doi.org/10.1007/978-3-319-76654-6_8

as heirs to their Jewish and Israeli past as described in Chapter 2, and the personal experiences they undergo as children and in young adulthood, which is explored in this chapter. Not all Israelis, most notably Israeli Arabs, take part in these shared experiences, but for the majority of those who do it is both intense and remarkably consistent. Israel as a society is small and close-knit by the standards of the modern world, and it imposes strong demands of its citizens in terms of national identification and community. This is, of course, a function of the state of war or near war that Israel has lived in since its creation. Thus, while family, friends, schools, and the media all play a role in the shared Israeli experience, the principal institution that forms the Israeli character is the army.

The Israeli character is critical to understanding the country's high-tech achievements because, in contrast to Silicon Valley and other technology clusters in the USA and Europe, its industry was not built by a self-selected group of outsiders attracted to an emerging or established cluster. Silicon Valley didn't become the world's leading technology cluster because people in Northern California showed more talent or ability for innovation or entrepreneurship than other Americans but because the region attracted people from other parts of the U.S and the world with those qualities as evidenced by the fact that 43.9% of all engineering and technology companies formed between 2006 and 2012 in Silicon Valley had at least one founder who was born outside the country.[1] Other global high-tech centers like London and Berlin similarly rely on a global workforce and even actively recruit it.[2] With about 25% of the population in 2015 born overseas, Israel has an unusually large number of first-generation immigrants. Moreover, immigrants to Israel, especially from the former Soviet Union, have an especially strong educational profile that would make them ideal candidates to start-up companies. But Israel's immigrants came for ideological reasons or to escape anti-Semitism, not as a self-selected group in pursuit of business opportunities in high tech. If anything, immigrants are underrepresented in Israeli high technology, with one survey finding that just 10% of the high-tech workforce had a degree from a foreign institute of higher education.[3]

Few visiting foreign entrepreneurs or engineers work in the Israeli tech sector either. Ironically, Israel is relatively liberal about issuing visas to guest workers in manual trades like agriculture, construction, and home healthcare. But for foreigners who want be a part of Israel's high-tech scene as an entrepreneur or engineer, the obstacles are daunting. Israel's "expert worker" visa program is very difficult to qualify for, even for someone the requisite skills, experience, and training.[4] The sponsoring employer needs to prove that the candidate's salary will be at least double the market average, demonstrate his or her expertise, and show that he or she has skills unavailable in Israel.[5] The restrictive visa policy was for many years a marginal issue because Israel never expected to attract large numbers of engineers due to the volatile security situation. But in recent years, the country's reputation as the "start-up nation" and Tel Aviv's emergence as a city offering an attractive lifestyle in terms of culture and entertainment, more foreigners have shown an interest and the companies facing a local labor shortage have shown an interest. Moreover, the scores of multinational research and development centers operating in Israel, which rely on an easy flow of staff moving from one country to another, are particularly hurt by the barriers.[6] In response, the government approved a plan late in 2015 for a two-year "innovation visa"[7] to be awarded to 50 foreign entrepreneurs with a declared interest in starting up a company. However, the program, which echoes others that have been introduced by countries like Australia, comes two decades after the start-up phenomenon took hold and will, given its small scale, have little impact on it going forward. Israeli high tech will remain a purely Israeli phenomenon for the foreseeable future.

The Israeliness of Israeli high tech is mainly due to the army, which serves as the principal conduit for transmitting the values that form the basis of the Israeli character to successive generations of young people. The critical place of the Israel Defense Forces in Israeli society is only partially captured by numbers. The size of the army's standing force is not publicly available but it is estimated to number about 177,000, which amounts to just 2.3% of Israel's population. Mandatory service is not especially long: Men currently serve 32 months and women 24 months, although those who pursue specializations or are promoted to non-commissioned officers serve somewhat longer. All of that would suggest that the army's impact on Israeli society is ephemeral. But that is not the case at all. The percentage of young Israelis conscripted is quite high, especially for the core population of Jewish Israelis (Israeli Arabs are exempt from mandatory service, although some non-Jewish minorities serve). In 1990, a good taking off point for the start-up phenomenon that arose in the following decade, 74.7% of all men of eligible age were drafted.[8] The figure has gradually fallen over the years to about 70% in 2000 and was about two thirds in 2012, but the decline is due mainly to the growth of Israel's non-core ultra-Orthodox population:

More than half of the male Jewish population that isn't drafted is from the ultra-Orthodox community, which is generally exempt from mandatory service. Inside the core population of young Israeli males, the conscription rate remains very high. Even among Israeli-Jewish women, close to 60% are drafted, and nearly all the women who don't serve were exempted for religious reasons, which for women includes not only the ultra-Orthodox but those with less strict standards of religious observance. Not counting the exempted populations of Israeli Arabs and the ultra-Orthodox, some 60% of Israelis have at one time in their life served in the army.[9] Moreover, the IDF's impact continues long after conscription. Most men continue doing reserve duty for up to one month a year, in some case up to their mid-forties, while those who go on to become career officers are typically cashiered while still in their prime working years and move on to key positions in industry and government. The army's impact is even greater in the high-tech industry, where one 2013 survey found that army veterans accounted 89% of all people employed.[10]

The IDF's role as an incubator for future technology entrepreneurs and engineers is usually attributed to its lauded technology units, most famously the 8200 intelligence corps, but that only partly explains the strong link between army service and Israel's start-up culture. Even if they do not teach engineering skills, the army's combat units are a major contributor to the Israeli national character and its high-tech sector by helping to create an ethos of entrepreneurship and innovation. Thus, while some 31% of Israeli high-tech personnel served in a technology unit, a nearly equal 29% served in a combat unit.[11] Among the start-up companies, the rates were even higher, with 36% having served in a technology unit and 32% in a combat unit. This link between the army and high tech would at first glance seem to be counterintuitive: Armies, even in democratic societies, are by nature big organizations subject to rigid hierarchies. They are bureaucratic and conservative, especially in peacetime when the urgency of the battlefield gives way to routines and red tape. Even though armies welcome new technologies, they are principally valued for their cost and firepower; when innovation (as it often does) threatens military norms and traditions, armed forces are generally less receptive to it. The Israeli army certainly shares some of these characteristics, but because of its unusual role in Israeli society and the fact that it faces constant, immediate security threats and frequently fights actual wars, it retains a battlefield ethos and an unusual level of nimbleness compared to other armed forces that radiates down to the level of the ordinary soldier. A Rand Corp. study comparing the American and other militaries noted that the Israeli army was more "adaptable" to public criticism (civilian commissions investigating the army's operations have followed perceived military features since the 1973 Yom Kippur War) and responsive to change than the US Army.[12]

The unusual role of the IDF goes back to the 1950s and 1960s, when Israel was a young state still in the process of actively creating a national culture out of a population comprised mostly of recent immigrants. The army was a natural candidate to play the role of melting pot. With Israel facing perpetual security threats and its population tiny compared to most of its enemies, the IDF enjoyed a pride of place as defender of the nascent state and conscription rates were high. The army brought a variegated population of Israelis together by subjecting them to the same training and routines, breaking down the values and traditions they came with through the discipline and the rigors of basic training as well as through an intensive program of educating recruits in the proclaimed socialist values of the state, as the establishment saw it. Apart from Zionistic patriotism, these values emphasized equality, self-sacrifice, dedication, and identification with the group and at the same time personal initiative. Some of these values are formal and explicit, but many of the most important are tacit. Their origins lie deeply in the labor Zionist ethos described in Chapter 2, but their proximate origins can be traced to the "sabra" ideal that coalesced in pre-state Israel. Among these sabra values was a strong distaste for hierarchy, even in the military context, with its uniforms, medals, and ranks. The sabra ideal emphasized action over theory, the role of small, purpose-driven groups as against big organizations ("to be one of the guys, not just a cog in the military machine") as well as a culture of straight talk, shorn of formalities, and deference.[13] These values were absorbed by the Palmach, the elite fighting force of pre-state Israel, and later when the Palmach was absorbed into the newly formed IDF, dispersed throughout the army.[14]

The IDF's role as a melting pot for immigrants has receded in importance as the rate of immigration has fallen, but it retains its critical social role. For many young Israelis, army service is their first contact with a wide spectrum of other Israelis, the differences ground down by an all-enveloping army culture. Of course, schools, youth movements, and the media also play a role of creating the Israeli national character, but the IDF has the advantage of gathering a majority of the country's population into a single organization at an especially critical time of their lives when they are emerging into adulthood and away from their homes and the immediate influence of friends and families. Simultaneously, the army acts as a filter for placing young people in tasks according to their intelligence and abilities, much as standardized testing, high schools and universities do in other developed economies. These range from office work and semi-skilled jobs to an elite of top combat units, pilots, intelligence analysts, and technologists. This process of "military socialization" cultivates new skills, social networks and norms and behavior codes creating a kind of "military capital,"[15] which constitutes an important subset of the intellectual capital employs in its high-tech industry.

This process of military socialization can be broken down in four basic stages. The first is *gibush*, a pre-induction trial where young recruits, principally for elite combat units, are sent through a battery of physical and mental tests over a period of days to filter out the best candidates. Roughly translated as team-building, gibush involves recruits being assigned tasks of the kind you might encounter in the battlefield that involve group problem-solving (often, to their frustration, problems designed to be unsolvable) under the pressure of knowing they are being watched and judged. The tester is interested in whether the group can solve the problem but even more so in the group's dynamics—who conceives of a solution, who leads the group, who provides morale. The second stage is basic training, which all recruits go through to one degree or another, with future combat troops undergoing a lengthier process of several months and those doing clerical work as few as two weeks. IDF basic training does not differ appreciably from other armies, although in the Israeli context it marks a particularly traumatic—albeit brief—interruption in the egalitarian, informal, and communal social values that typify Israeli society and the army. Apart from teaching raw recruits combat skills, the critical aspect of the process is the strict regimen of discipline it imposes on recruits both in terms of hierarchical relations between the commanding officers and the micromanagement of the recruit's time, activities, and behavior. The first involves imposition of what is called in Hebrew *distans*, in which recruits must not only obey their commanding officers without question or hesitation but must refrain from any personal interactions with him or her, even as simple an act at laughing at a joke together. Distans aims at creating a sense of hierarchy and obedience in a society that has little of either. The second aspect of basic training aims to reduce the new recruits to an undifferentiated mass all engaged in the same tasks and held to the same standards, reinforcing the team ethos the army is trying to build on.

This phase ends relatively quickly and abruptly with an event called *shovrim distans* (breaking the distance) in which the officers meet their soldiers, reveal personal information about themselves and listen to soldiers' candid judgment of their performance. This, in effect, is the beginning of the real IDF, where discipline and hierarchy are by the standards of most armed forces quite lax. One young man, who had the unusual experience of being a raw recruit in both the Israeli and US armies, noted the difference. "Rank structure absolutely exists [in the IDF], but at the same time all enlisted ranks are more of less equal 90% of the time…when you're on a base and you're dealing with your commanders – up to a lieutenant, or even sometimes a captain – as long as you're not in a frontline combat unit where everything is a little more rigid – there's a good chance that within a few months you're going to be addressing him on a first name basis."[16] Edward Luttwak, an American political scientist and expert on military strategy, notes that the IDF has relatively few officers relative to enlisted personnel—a ratio of 1 to 9 versus 1.5 in the US Army and concludes that it is a deliberate policy designed to ensure more individual initiative among lower ranks.[17] In combat units, improvisation, especially in time of war, is the norm and is even expected of low-ranking soldiers. The ethos of individual initiative extends to the principle that disobeying orders does not automatically lead to punishment, if it was done with the goal of completing a mission or task.

Israel's egalitarian culture is certainly a factor in creating and perpetuating this culture, but it also reflects the paucity of manpower and materiel the army has to contend with.[18] As one example, the Israeli Air Force is large by world standards, with an estimated 243 combat aircraft,[19] but relative to the missions it has to fulfill and the enemy forces arrayed against it is modest in size. Because the resources at its disposal are simply too small for specialization, the IAF expects its planes to perform multiple tasks and its pilots to act with a degree of independence and self-management that larger air forces do not demand, in fact discourage. Thus, unlike the process for selecting pilots in the US Air Force, which focuses on aptitude for flying, the IAF pays equal, if not more, attention to character, motivation, and leadership traits.[20] "Beyond producing technical operators, leadership [ability] is scrutinized throughout the program. Initiative is inculcated and rewarded at all levels, and the warrior-leader ethos permeates every facet of the IAF system."[21]

The next component for many young recruits is an unusually intense and rigorous training period that prepares them for their army roles and soon thereafter gives them enormous professional and command responsibilities while they are still in their teens and early twenties. Whether it is Arabic language or computer science, this training occurs for almost all recruits before they enter university and is very different from what they would receive in an academic setting. It is completed in the space of a few months, enveloping young soldiers in whatever course work they are involved, imparting unusually deep skills, and emphasizing problem-solving over theory. One graduate of the 8200 course describes a typical classroom session in which young recruits learn how to produce intelligence, leverage the most advanced SIGINT (signal intelligence), utilize sophisticated data-mining techniques, and conceive highly advanced technologies[22]:

> On [a] particular rainy day, our instructors ran us through a simulation exercise.
>
> They provided us with hundreds of short, fictional pieces of intelligence. Each one, on its own, appeared inconsequential. Very quickly, however, one of my classmates, a future intelligence officer, began to piece together the puzzle presented to us.
>
> He yelled: 'A war is about to break out!!'
>
> An intense debate erupted among the trainees about the true meaning of the seemingly unrelated information we had been provided.
>
> Our instructors had used the simulation to stimulate a heated discussion and, perhaps more importantly, a leadership test case. While we were passionately arguing whether a war was about to take place in our fictional state, our instructors dramatically stopped the simulation and ended the discussion. They told us that the simulation was based on real-life events, and indeed, a war had broken out. My classmate had been right.

"It was an important, poignant lesson for my classmates and me. We learned that succeeding in intelligence work required more than just discipline and professionalism. Success required out-of-the-box thinking, the courage to contradict conventional wisdom, and an ability to stave off hubris. A good intelligence officer needed to understand when to bypass hierarchies and be willing to take risks and make mistakes."

Soldiers entering elite combat units as well as pilots and those serving on submarines go through advanced training to teach them the skills relevant to their tasks, but the ethos is the same, namely not simply to learn the required technical skills but to develop their analytical and decision-making abilities and to use them inside a team environment. Soon after, they will be leading complex intelligence-technological operations or commanding troops, so that their intense training in those skills is employed not just in the classroom but in real-life conditions. As the 8200 recruit explained, "Instead of relying on outside research and development, the 8200's technologists work directly with their 'customers' (the intelligence officers). All of the unit's technology systems, from analytics to data mining, intercept, and intelligence management, are designed and built in-house. Technologists sit side by side with their users on a daily basis to ensure that their 'products' meet the intelligence officers' specific requirements."[23] The downside of this kind of the military human capital soldiers bring to the high-tech sector is a relative disregard for a business way of thinking that takes into account cost or time to market. In that way, army R&D is similar to academic research, which is the second principal source of basic R&D in Israel. Unlike the army's mission orientation, academe emphasizes research for the sake of general knowledge rather than for applications, but it shares the same absence of business thinking, which is passed on the country's high-tech sector.[24]

This army experience, acquired in combat and technology units alike, contributes four key national characteristics that have been responsible for Israel's start-up culture—and at the same time hold it back from capitalizing on its innovative capabilities.

Distaste for organization, hierarchy, and rules: Israelis dislike and resist the structural and disciplinary demands of large organizations. This might not appear self-evident because the IDF itself is a large organization capable of putting tens of thousands of troops into battle and overseeing large-scale technology projects with budgets in the hundreds of millions of dollars, but its successes in the battlefield and in developing defense technology are less a function of efficient organization than personal and team initiative taken in the face of immediate security needs. "Israel is capable of bringing to fruition projects like [the anti-missile system] Iron Dome and units like 8200 turn out large-scale technology, but the motivations are different – they are based on national security considerations, or more directly national existence – your back to the wall," says Sigal Widman, a strategic and organizational consultant for the Israeli high-tech industry.[25] Israelis work best in small groups assigned a specific goal, which in business means they work well as teams of start-up entrepreneurs or, in bigger businesses, as research and development teams assigned to develop an innovative, new product. Ideas are aired freely and equally even when they come from younger and inexperienced members of the group, while higher-ranking member can expect little deference. Issues are analyzed, criticized, and amended in a process unconstrained by rank, authority, or organizational processes. Where Israelis are far weaker is operating inside the large organizations where lines of authority, processes, planning, and long-term strategy are critical in creating a sustainable business. Thus, once the innovative phase of a project or nascent start-up company is complete, the discipline required of corporate organization becomes paramount, and Israelis are more likely to fail.

The Israeli attitude is captured in a story that has taken on legendary status in Israeli high tech about Israeli engineers' campaign inside Intel Corp. to break with the company's decades old business and technology strategy in chip development.[26] Since it was formed in the 1960s, Intel, like the rest of the industry, measured advances in semiconductor development by clock speed, the rate at which chips perform functions. But by the end of the 1990s, faster speeds were running up against a technological barrier: As chips became smaller and faster, they produced so much heat that they needed increasingly bigger fans to cool them — so big that they could not be put into smaller, thinner configurations of the emerging generation of notebook computers. Intel's Israeli research and development team was the first inside the company to recognize the problem and began to pursue solutions informally and then eventually was put in charge of mobility chips. The solution the team presented to Intel's managers was a chip that could run software faster but used less power and generated less heat by splitting the instructions fed into the chip. Headquarters rejected the idea because, by the industry benchmark of clock speeds, the chip was slower and a step backward. Paul Otellini, then head of the company's chip division, wanted to drop the idea altogether. Rather than accept management's dictate, the Israeli team persisted in making its case, despite its being a small and distant outpost of a giant multinational company. "We did it the Israeli way; we argued our case to death," recalled Shmuel (Mooli) Eden, who has head of Intel's Israel Development Center at the time. "You know what an exchange of opinions is in Israel? You come to the meeting with your opinion, and you leave with mine."[27] The Israeli design was finally accepted and introduced in 2003 as Intel's immensely successful Centrino product.

Risk-taking: By its nature, a start-up company is a risky venture even by the standards of new businesses. Entrepreneurs risk their capital, reputations, and (most importantly in the high-tech industry) potential earnings that they could easily enjoy working for an established company. In fact, the risk is greater for technology entrepreneurs than other nascent entrepreneurs. The latter typically enters an established market, exploiting a perceived opportunity, and organizes his or her new enterprise according to standard business principles as they relate to costs, marketing, and the like. By contrast, a technology entrepreneur starts his or her enterprise with an untested idea in a market that may not yet exist or, if it does, is undergoing rapid and unpredictable change. Thus, even in Israel, where the start-up industry is highly institutionalized in the form of accessible venture capital, angel investors, serial entrepreneurs, investment bankers, technology accelerators, and training programs, all of which serve to lower the risk involved, only between 2.5% and 6% of all start-ups succeed, depending on the criteria used.[28]

The vastly different risk profile might explain why economies where traditional entrepreneurship is widespread, high-tech entrepreneurship isn't necessarily prevalent. A survey Israeli small-business owners versus high-tech entrepreneurs confirms this by showing considerable differences not only in the educational and social background of the two groups but also in personality traits. High-tech entrepreneurs were better educated and far more likely to have served as an army officer, confirming the army-tech linkage.[29] Small-business owners were significantly more likely to describe themselves as more realistic, as preferring to manage and exercise control than tech entrepreneurs; tech entrepreneurs, by contrast, tended to describe themselves as greater dreamers, risk-takers, lovers of challenges, and as more creative.[30] "Entrepreneurism at the end of the day is driven by individuals. You have someone who has decided to challenge the status quo in some industry elsewhere," says Michael Eisenberg, a venture capitalist, explaining the kind of personality he looks when weighing investing in a start-up. "It's a one in a million business, so you're looking for a one in a million entrepreneur, really smart, really successful, real disruptive, who will shake up an industry."[31]

Israelis' high propensity for risk-taking is partly a function of the uncertainty that has characterized Jewish life for centuries. In Israel, much of this legacy is captured in the memory of the Holocaust, a subject that receives much attention in schools and in public life. By itself, historical memory would probably not be enough, especially as the generation of survivors grows old and dies, but the uncertainty is kept alive and present because of the country's perpetual war-footing and the anxieties about national survival it creates. That anxiety is evidenced in a study that demonstrates the impact of security risk arising from terrorism on economic growth during the first 55 years of Israel's history. Not surprisingly, growth was higher in periods when risks were relatively low (most notably in the 1951–1973 and 1989–2000 periods).[32] During periods of higher risk increases in defense spending accounted for some of the loss of economic growth, but the study also attributed it to "an increase in uncertainty about life" on the part of consumers, investors, and business people.

This is less self-evident than it might appear because by many relevant social measures, such as life expectancy and access to health services, Israel matches the world's most advanced economies. But Israeli anxiety about the risk of terror is unrelated to actual outcomes. The periods of high risk were not characterized by lengthy wars or major losses of men or materiel, with the exception of the 1973 Yom Kippur War. Even during the wave of terror that characterized the first years of the Second Intifada from 2000, the rate of deaths from attacks was about the same as from automobile accidents. Yet, fear of death or injury by terror occupies a special place in economic considerations. Its unpredictability and the fact that it receives more media attention than ordinary deaths and injuries serve to ensure a place for it deep in the public conscious. In turn, that influences the decisions of entrepreneurs about the kinds of businesses they will choose to enter and what resources they will put into it. Israelis value life (indeed they arguably put an unusually high value on it), but they are acutely aware about its tenuousness—an attitude that inevitably extends to their perception of the world as laden with danger while at the same time inuring them to the danger. One small example of that is a survey taken in the autumn of 2015, amid a wave of stabbing attacks by Palestinians of Israeli civilians in public places. The poll showed that 57% of Jews feared for their safety or that of loved ones but at the same found that nearly two thirds reported they had not changed their routines by refraining from public transportation less or altering their shopping habits.[33] In a life perceived to be laden with risk, the speculative nature of start-up entrepreneurship doesn't seem out of place.

Security fears also express themselves in the economy's disinclination to invest in capital goods and focus more on human capital, which is portable and less likely to be lost in conflict. In a study of Israel's low level of labor productivity, the Bank of Israel found that for the years

2000–2011 investment in fixed assets—that is land, buildings, and factory equipment—was just 17% of gross domestic product, compared with 22% for other developed countries, even though the return on capital for most of that period was 25%, two percentage points higher than the average for the Organization for Economic Cooperation and Development.[34] The study suggests that the security risk coming with a history of wars and terror attacks deters investment in physical assets, although it should be noted it did nothing to deter Israelis from obtaining a higher education or starting up companies. In this context it is easy to understand how high-tech start-ups, which typically have few physical assets and invest principally in human resources, are an attractive business for Israelis. However, the downside of this disinclination to invest in capital goods is that it erects another barrier to Israelis' ability to build large, sustainable enterprises.

Culture of teamwork and group loyalty: The gibush process is not unique to the army; the idea that any group you belong to must establish a dynamic and create a sense of group spirit pervades Israeli society from elementary school, where teachers seek to establish gibush on the classroom, to the workplace, where employers often try to do the same. David Galanti, who was head of operations is Israel and head of wireless research for the American chip maker Freescale, says the different in attitudes in terms of team loyalty can be observed in how frequently Israelis change jobs, compared to their counterparts of China and India. In a 2012 interview, he estimated that turnover of engineers in China and India is 20–30% a year and staff stays with the same employer for an average of two years.[35] "It's enough that they get an offer for more money or to work on a more interesting project to cause them to move to another company. In Israel, changing workplaces is done mainly on the basis of long-term career considerations, professional advancement and the character of the company," he says. Teamwork and group loyalty are critical in innovation. Conventional wisdom sees innovation as the result of individual insight, creativity, and labor but in high technology that is very rarely the case. Innovation is typically the product of shared thinking, often across engineering and market disciplines that require more expertise than one individual can master.

An important element of this team approach in Israel is that it doesn't foster group think. This relates to an important Israeli cultural characteristic, which values candidness and directness often to the point of what outsiders can perceive as rudeness. The group functions not to impose a certain point of view, rather as a forum for exchanging ideas and opinions, its members held together not by shared opinions but by shared experience and personal intimacy, reinforced by a culture that encourages direct language and outspokenness. David (Dadi) Perlmutter is quoted as recalling a meeting of Intel Israel staff, with an American executive in attendance. "When we all emerged [from the meeting] red faced, after shouting, he asked me what was wrong. I told him, 'Nothing. We reached some good conclusions.'"[36] This culture of directness and debate is encouraged by the army but is also deeply embedded in Jewish life. The Talmud, which among Orthodox Jews has more influence on religious life than the Bible, is principally a record of rabbinic debates over matters of law and philosophy, expressed in condensed language and often times with the conclusions left unresolved. In the pre-state Israel, the culture of debate moved from the religious to the political sphere and was fortified by the Sabra generation whose values infuse so much of Israeli culture. Sabra language, developed in a military framework, was informal and impolite and lacked and protocol for acknowledging authority or rank.[37]

Problem-solving/task orientation: The world perceives Israel as a regional military power and an advanced economy, its place in the world community an established fact. But the great majority of Israelis continue to see the state's survival as perpetually in jeopardy and every conflict it engages as a fight for its existence. Former Prime Minister Levi Eshkol called this phenomenon of objective strength and subjective weakness and doubt "Shimshon der Nebedicher" (Samson the Nebbish). Like other Israeli characteristics, it is a product of the precariousness of Jewish life through centuries of diaspora and the strategic threats the country faces, as well as the paucity of human and material resources Israel can call on to meet challenges. "The diaspora created a culture of organizing on the fly because [Jews] were perpetually in a situation of chaos and uncertainty. Jews and now Israelis created a culture of implementational ability as seen even in the formation of the State of Israel itself – in a situation of war and violence – surrounded by enemies – which demanded immediate and creative solutions," says Widman.[38] Even enormous projects involving hundreds or thousands of people and budgets in the hundreds of millions of dollars that would be framed in other societies in organizational terms, in Israel are goal directed, the motivation being an immediate identifiable threat or problem that must be overcome inside a very constrained timetable. This is not to say that the IDF is not a large and bureaucratic organization, or that organizational demands play no role in its decisions, but they are less central to motivating the people involved.

The Iron Dome anti-missile system, a major military project, captures virtually all four of these Israeli characteristics in a single story. The system was designed to identify and intercept primitive short-range rockets flying irregular trajectories that reach their targets within tens of seconds after launching—an especially difficult task that no other anti-missile defense system could provide. Israel had little choice but to devise a solution to the problem and to do it quickly because the country's home front faced the threat of rocket barrages from the militant groups Hezbollah in Lebanon and Hamas in the Gaza Strip. The size and scope of that threat became especially evident after the 2006 Lebanon war when Israel was pummeled by thousands of rockets by Hezbollah and subsequently on a smaller scale by Hamas.

The US defense establishment rejected the idea in principle that any anti-missile technology could succeed against such targets, as did much of the IDF leadership.[39] Nevertheless, Brig. Gen. Daniel Gold, who was director of the Defense Ministry's Research and Development Department, was convinced that the problem was solvable and moved forward with the project on his personal initiative. His team examined a range of technology options before deciding on a system pieced together from technologies developed by three different Israeli defense companies and then went around usual procedures for defense contracting to assign the work to one of them, state-owned Rafael. Gold was criticized by the government's State Comptroller's office for the irregularities, but he was never disciplined by the army. Iron Dome, which took four years to develop (less than half the time it usually takes for similar projects), proved to be a successful and relatively low-cost technology for intercepting rockets. "I just canceled all the unnecessary bureaucracy," Gold told *The Wall Street Journal* later. "I left only the most crucial bureaucracy needed for success." Thus, while Iron Dome entailed massive organization and funding, it was not a typical defense project but more akin to World War II-era Liberty Ships, a method of rapidly building cargo ships developed quickly by British and American engineers under the pressure to constantly replenish a merchant marine being decimated by German U-boats.

In a knowledge economy, an effective educational system, developed infrastructure, liberal and tolerant social attitudes, and appropriate rules and regulations are all important factors in exploiting the potential of its human capital, but in Israel's case, the unusually high intellectual capabilities of the population as measured by IQ almost certainly play a critical role as well. The issue of IQ is enmeshed in controversy over its social and political implications and because practical reliance on it as a measure of human potential risks sliding into racial and ethnic bias. But the fact is that IQ tests are a very good predictor of future success in academic work and jobs requiring cognitive ability, which is the foundation of any knowledge-based economy. In this context, Ashkenazi Jews have been shown to have the highest average IQ of any ethnic group for which there are reliable data, corresponding to a range of 112–115. Not just in Israel, but globally they have demonstrated extraordinarily high levels of achievement both by broad measures such as education and average income as well much narrower ones such as Nobel Prizes.[40] This high-IQ distinction belongs solely to Ashkenazi Jews, who trace their origin to Europe and North America, as against Sephardi and other Jews who until most immigrated for Israel or the West in recent decades lived in the Middle East and North Africa.

What percentage of Israel's Jewish population is "Ashkenazi" is impossible to say, not only because census data don't categorize people by that definition but because the category itself is fluid and the rate of intermarriage among Jewish Israelis means many have mixed parentage sometimes going back two or more generations. But certainly Ashkenazi-origin Israelis constitute a majority of Israeli Jews[41] and they have brought their capabilities to bear on the country's economic and cultural life. The success of the Israeli high-tech sector, which is characterized first and foremost by its capacity for innovation as against other business-related skills, is almost certainly connected with Ashkenazi Jews' high levels of verbal and mathematical intelligence (although, interestingly, while Ashkenazi Jews perform exceptionally well in tests of verbal and mathematical ability, on average they score lower on visuospatial abilities than Europeans).

The origins of the high-IQ phenomenon go back quite possibly to the Second Temple era, starting around 100 BCE and the gradual rise of rabbinic Judaism in place of the temple cult. The focus of Jewish practice shifted in the next two centuries shifted from temple worship in Jerusalem to community-based prayer and study, which required that Jewish males be literate. In turn, that led to the establishment of local schools, producing unusually high rates of literacy among Jews in Palestine and the diaspora at a time when schooling and the ability to read and write was the preserve of a tiny minority of the very wealthy and a small class of scholars. Maristella Botticini and Zvi Ecvkstein suggest that the burden of the schooling requirement was too much to bear financially for many Jews, who in this era were mostly peasants for whom education generated no economic returns.[42] In the first centuries of the Common Era, an era of economic stagnation in the Mediterranean era, many Jews opted to convert, in many cases to nascent Christianity. The creation of the Muslim empire in the seventh century CE revived economic growth and urban culture, causing Jews to settle in the cities and take skilled, cognitively demanding occupations crafts, trades, money lending, tax-farming, and medicine.[43] The limited number of jobs of these kinds in a pre-modern economy was limited, so in order to maximize the return on their educational investment, they migrated within the Muslim world and from the ninth century on to Europe as well.[44]

In Europe, the emergent Jewish population was tiny and specialized in similarly high-cognitive occupations as financiers, estate managers, tax farmers, and merchants, initially trading between the Christian and Muslim worlds and eventually engaging in local commerce.[45] Gregory Cochran, Jason Hardy, and Henry Harpending suggest that the process of rising IQ among Jews occurred in Europe between ninth and sixteenth centuries.[46] The tendency to engage in high-cognitive occupations encouraged them to continue investing in education, even if the proximate motivation was religious not economic. Jews as a rule did not intermarry, so there was little gene inflow from outside the community, while the rich tended to have more children than the poor.[47]

In the Middle East, the phenomenon of Jewish specialization in cognitively demanding professions came to an abrupt halt with the Mongol invasions from 1250 on, which brought the destruction of many of the Muslim world's biggest urban centers and a rapid decline in population.[48] In the absence of occupations that could justify the cost of educating their children, over the next two centuries large numbers of Jews converted to Islam, which made no such demands.[49] In Europe, meanwhile, Jews eventually lost control of the most cognitively demanding occupations to gentiles. Many migrated to Eastern Europe in the late Middle Ages where their skills continued to be put to use, but by the seventeenth century, the Jewish population was too big for the majority to be engaged in them, and Jews began drifting to lower-skilled occupations. By then, however, the process of IQ development had already occurred.[50]

The intangible element of Israel's human capital asset base has created fertile ground for the rise of the country's high-tech sector, but it also creates barriers to the dispersion of these knowledge assets to the wider economy. These intangibles that are so critical to Israel's start-up environment are at the same time antithetical to these needs and values of large organizations, which means that Israel's innovative strengths prevent it from capitalizing on them by creating and sustaining large enterprises. These intangibles are similarly ill-suited to industries where rules and regulations are preponderant, which explains to a large extent why Israel has failed to win itself a place as a global financial center. Thus, Israel's acknowledged strengths in financial technology (fintech)[51] show no signs of being leveraged into creating world-class financial institutions. Finally, it is the nature of Israel's human capital intangibles that there is an upper limit to how much Israeli society can expand and enhance them to a wider swathe of the population. As discussed in Chapters 10 and 11, Israel's Haredi and Arab minorities, the country's main untapped reserves of human capital, largely exist outside the framework where these intangibles are rooted and cultivated. Even if Israel succeeds in ensuring equal access to formal education and job opportunities, the process of having these minorities embrace these intangibles will take time and, indeed, may be impossible.

Notes

1. Vivek Wadhwa et al., *Then and Now: America's New Immigrant Entrepreneurs, Part VII* (Kansas City, MO, October 2012), 25. Indeed, the USA as a country serves as a magnet for the world's technology entrepreneurs, with 24.3% of all tech start-ups across the US counting at least one non-American founder.
2. Mark Scott, "Technology Start-Ups Take Root in Berlin," *New York Times*, April 29, 2013; and Murad Ahmed, "Foreigners at the Helm of One in Five UK Tech Groups," *Financial Times*, March 28, 2016.
3. Orr Hirschauge, "Profile of Israeli High Tech: A Young Male, Veteran of an IDF Combat or Tech Unit and a University Graduate" (Hebrew), *TheMarker*, March 11, 2013.
4. Amir Mizroch and Orr Hirschauge, "Foreign Entrepreneurs Face Israeli Immigration Hurdles," *The Wall Street Journal*, September 23, 2014.
5. Iftach Hanan, Hi-Immigration Law Office, www.en.hi-immigrationlaw.com/Work-visa-in-Israel-Foreign-Experts.html; and John Reed, "Israeli Entrepreneurs and Immigration: 'Start-Up Nation' Stifled," *Financial Times*, November 11, 2014.

6. Amir Teig and Inbal Orpaz, "I Have 40 High-Paying Jobs That I Can't Fill," *TheMarker*, November 7, 2013.
7. Meir Orbach, "Economy Ministry Approves Grating Visa for High Tech Entrepreneurs from Overseas" (Hebrew), *Calcalist*, October 20, 2015.
8. Gili Cohen, "IDF Statistics: One in Six Israeli Men Doesn't Complete Full Military Service," *Haaretz*, June 9, 2013. The latest figures are for 2012.
9. Ori Swed and John Sibley Butler, "Military Capital in the Israeli Hi-Tech Industry," *Armed Forces & Society*, Vol. 41, No. 1 (January 2015), 125.
10. Orr Hirschauge, "Profile of Israeli High Tech."
11. Ibid.
12. David E. Johnson et al., *Preparing and Training for the Full Spectrum of Military Challenges: Insights from the Experiences of China, France, the United Kingdom, India, and Israel* (Santa Monica, CA, 2009), 231.
13. Oz Almog, *The Sabra: The Creation of the New Jew* (Berkeley, CA, 2000), 104, 145, 218–219, 228, and 245.
14. Ibid., 132–133, 231.
15. Swed and Butler, 124.
16. Mark Thompson, "One Private, Two Armies," *Time*, February 13, 2012.
17. Dan Senor and Saul Singer, *Start-Up Nation*: The Story of Israel's Economic Miracle (New York and Boston, 2009), 60.
18. Ibid., 59.
19. Craig Hoyle et al., "World Air Forces: 2016," *Flight International*, December 3, 2015, 11–34. A few examples of other major world air forces: The USA counts 1,920 combat aircraft, Germany 169, France 226, and Egypt 427.
20. Michael D. Hays, "The Training of Military Pilots: Men, Machines and Methods," June 2002, 49–50.
21. Ibid., 51.
22. Ilan Tendler, "From the Israeli Army Unit 8200 to Silicon Valley," *TechCrunch*, March 20, 2015. Tendler was at the time of the article CEO of Fortscale, a network security start-up.
23. Ibid.
24. Dan Brenitz, *Innovation and State: Political Choice and Strategies for Growth in Israel, Taiwan and Ireland* (New Haven, CT, 2007), 42–43.
25. Interview with Sigal Widman, July 31, 2012.
26. Various elements of this account are taken from: Ian King, "Intel's Israelis Make Chip to Rescue Company From Profit Plunge," *Bloomberg News*, March 27, 2007; and Dan Senor and Saul Singer, *Start-Up Nation*, 38–46.
27. King, "Intel's Israelis Make Chip to Rescue Company from Profit Plunge."

28. Israel Venture Capital Research Center and Reversexit, "Only Four of 100 Start-Ups Succeed; Only Four of 500 are Successful Growing Independently," January 28, 2015. The survey covered approximately 10,000 companies formed between 1999 and 2014. The narrow criteria for success were companies with at least 100 employees, annual revenues of at least $100 million, or acquired for more than the total invested in the company. The wider criteria were companies with more than 50 employees, fund-raisings at valuations of at least $50 million or have operated with raising capital in the previous three years.
29. Dafna Schwartz and Ayala Malach-Pines, "High Technology Entrepreneurs Versus Small Business Owners in Israel," *Journal of Entrepreneurship*, Vol. 16, No. 1 (2007), 6. The respondents were asked to compare their personality traits to their parents.
30. Ibid., 8.
31. Interview with Michael Eisenberg, Benchmark Capital, April 15, 2012.
32. Zvi Eckstein and Daniel Tsiddon, "Macroeconomic Consequences of Terror: Theory and the Case of Israel," December 10, 2003.
33. Tel Aviv University and the Israel Democracy Institute, The Peace Index: October 2015, November 5, 2015.
34. Bank of Israel, *Annual Report, 2012* (Jerusalem, April 2, 2013), 51–53. In fact, investment in manufacturing is relatively high in Israel even if the manufacturing sector is relatively small. This, however, may reflect the fact that the government subsidizes large capital spending projects in manufacturing.
35. Ora Coren, "Today It's Clear That India and China Are Not an Immediate Threat to Israel" (Hebrew), *TheMarker*, July, 11, 2012, 20.
36. Senor and Singer, *Start-Up Nation*, 47.
37. Almog, 245.
38. Widman interview.
39. The history of Iron Dome's development comes largely from Charles Levinson and Adam Entous, "Israel's Iron Dome Defense Battled to Get Off Ground," *The Wall Street Journal*, November 26, 2012.
40. Gregory Cochran et al., "Natural History of Ashkenazi Intelligence," *Journal of Biosocial Science*, Vol. 38, No. 5 (2006), 659–693, 3.
41. The Central Bureau of Statistics reports that in 2014 of approximately 3.5 million Jewish Israelis either born abroad or whose father was born abroad, 55% traced their origins to Europe or North America, 25.5% to North Africa, and 19.4% to Asia (the Middle East). That group, in turn, accounted for 57% of all Israelis.
42. Maristella Botticini and Zvi Eckstein, "From Farmers to Merchants, Voluntary Conversions and Diaspora: A Human Capital Interpretation of Jewish History," *Journal of the European Economic Association*, Vol. 5,

No. 5 (2007), 885, 12–13. The thesis was expanded into a book, *The Chosen Few: How Education Shaped Jewish History, 70–1492* (Princeton, NJ, 2012).

43. Ibid., 26.
44. Ibid., 31.
45. Cochran et al., 6, 10 and 12.
46. Ibid., 2.
47. Ibid., 12.
48. Botticini and Eckstein 36.
49. Ibid., 39.
50. Cochran et al., 12.
51. Roy Keidar, "Israel: The Rise of the FinTech Hybrid," *Financier Worldwide*, July 2016.

CHAPTER 9

Inequality

Israel was founded on egalitarian principles, and they remain a powerful social value long after the country adapted free-market attitudes about the importance of competition, a flexible labor market and smaller government. Outwardly, Israel evidences little of the huge disparities in wealth readily apparent in other developed and undeveloped economies. The wealthy don't ostentatiously display their money with huge homes, luxury cars, and yachts. At the other end of the income scale, the social pathologies that typically accompany poverty, such as high rates of crime, teenage pregnancies, and broken families, are less frequent, to a large extent because many of Israel's poorest belong to traditionalist communities that impose behavioral constraints on their members, a subject to be discussed in Chapters 10 and 11. But the fact is that income inequality and poverty have grown sharply over the last two decades for a variety of reasons, some of them related to the same global phenomena that have affected nearly all developed economies and others particular to the Israeli experience. If there have been some signs more recently that income inequality and poverty are beginning to decline, they are too small and tentative to signal any reversal of long-term trends.

The rise of Israel's knowledge economy after 1990 came parallel with the increase in inequality and poverty. Although the two events were not directly linked, the knowledge economy, especially in the Israeli version as an industry of high-technology start-ups, certainly played a role in exacerbating the phenomena. The issue is not simply about social justice. Going forward, the future of the country's economy, both vis a vis the start-up sector

D. Rosenberg, *Israel's Technology Economy*, Middle East in Focus,
https://doi.org/10.1007/978-3-319-76654-6_9

and the ability to broaden knowledge-economy characteristics to encompass other segments of the economy, is tied up with the ability of policymakers to ensure greater income equality and make use of human resources lost because so many families are consigned to a life of poverty and all the disadvantages that entails. The Organization for Economic Cooperation and Development and the International Monetary Fund have both made the case for reducing inequality as a driver for long-term economic growth[1] because inequality by its nature diminishes an economy's ability to exploit its best and brightest by denying them adequate access to education, technology, and jobs. In Israel's case, ensuring greater income equality is especially critical—tantamount to an investment in the future—because its small population requires it to make the most of the human capital it has.

Inequality can be measured in several ways, but nearly all of them show Israeli society to be in economic terms among the least equal in the developed world. The most common indicator, the Gini coefficient, measures the distribution of income in households on a scale of 0 for complete equality to 1 for complete inequality, taking into account not just earnings generated by employment and investments but also the impact of taxes and social security contributions on the one side and government allowances on the other. By that measure, Israel in 2012 scored 0.371, the fourth highest among OECD countries after the USA, Turkey, and Mexico.[2] Since the latter two are not normally counted as wealthy economies, Israel and the USA were outliers in the developed world along with Britain, which was just behind Israel with a score of 0.351. The OECD average was 0.32, and the most equal countries in terms of income had scores of 0.25 or under. Looking at the gap between the top 10% of households by income versus the bottom 10%, again Israel shows a much higher level of inequality than its developed-country peers, the USA excluded. In 2012, the top 10% had on average 13.7 times the income of the bottom 10%.[3] The US ratio was 17.3, putting it in a category by itself among developed economies. But the Israeli ratio was still higher by a wide margin than any of the other developed economies surveyed by the OECD; the 10 countries with the smallest gaps all had ratios of 6.6 or less.

In a dynamic and competitive economy, a certain amount of inequality is inevitable (as people with differing abilities and levels of personal initiative will earn and save more or less) and even desirable (as a means of encouraging people to perform by rewarding them accordingly). But inequality feeds on itself by undermining the principle of equal access to the institutional resources that enable each and every member of the next generation to exercise his or her abilities to their fullest. Education is the most obvious area where this comes into play. Israeli workers with the least schooling made about 60% of the average wage in 2011 while those with 16 or more years earned 144%.[4] A gap like that is inevitable in a modern economy that puts a premium on skills and education. But it is clear that educational performance is strongly skewed by socioeconomic status. Using the results of the OECD's Pisa examination of educational achievement, researchers concluded that socioeconomic difference accounted for more than 15% of Israeli students' school performance in 2012, slightly higher than the OECD average.[5] The Israeli Education Ministry Meitzav examination shows how finely socioeconomic inequality affects school performance. Even as all groups showed improving performance over the years 2006–2014, students from high-income families consistently scored the highest for subjects like Hebrew language, math, science, and English, those from middle-income families had middling scores, and those from the lowest the worst performance.[6]

Unequal access to resources isn't limited to the schools. In a knowledge economy, familiarity with computers and the Internet is a sin qua non just for education and work but increasingly for access to government and business services, staying abreast of news and cultural events and even maintaining personal and social connections. But among lower-income groups in Israel technology use is much lower than for others. Among Israel's poor, personal computer ownership was a relatively high 55%, compared to the 84% for all other income groups, but the rate of Internet connection was just 39% versus 78% for all others—a 16 percentage point gap with PC ownership versus just six points for other households,[7] meaning a good percentage of the PCs in poor families were not used for their most critical function. A survey of Internet and PC usage confirmed that. The lowest income Israelis not only had the lowest rate of Internet connections but unusually high rates of people who didn't make use of the connection they had—37% of all those surveyed versus single-digit percentages for upper-middle and high-income groups.[8]

Meanwhile, in the brick-and-mortar world, there is evidence that upper- and lower-income groups in Israel have increasingly less direct contact with one another because more and more people live in separate towns and neighborhoods. The percentage of high-income households in exclusively high-income neighborhoods nearly doubled in Israel from 7% in 1983 to 13% in 2008. At the other end of the spectrum, the proportion of lower-income households in exclusively lower-income neighborhoods rose from 19% in 1983 to 30% in 2008.[9] Inevitably, that undermines the educational performance of students from lower-income families by consigning them to schools with fewer financial and teaching resources, or a critical mass of families with strong educational and socioeconomic backgrounds. More than that, it deprives them of facilities like parks, community centers, and youth groups that wealthier towns and neighborhoods typically enjoy because they have a stronger tax base and a more politically astute population. In short, the ability to rise out of poverty is increasingly constrained by the environment poor children begin their lives.

The Israeli army acts as a powerful force for creating social cohesion (as was discussed in Chapter 8), but it is less of a leveling tool for income equality. The army recruits young people for its top combat and technology units by employing testing and other scientific criteria, a system that not only serves to identify the best and brightest but also serves to perpetuate the differences that young people have inherited from their family, neighborhood, and school environments. A rough measure of the extent to which the Israel Defense Forces' elite units favor those from stronger socioeconomic backgrounds is evidenced by the geographical distribution of recruits. The greater Tel Aviv area, which is home to Israel's wealthiest families, accounts for 43% of all male recruits, but it makes up 54% of the army's software developers and those admitted to army academic programs.[10] By contrast, recruits coming from Israel's geographical periphery, which is generally characterized by poorer socioeconomic conditions, account for far fewer. About 15% of all recruits come from Israel's north, but they constitute for only 10% of software developers and those in academic programs. The gap for recruits from the south is 17% versus 14%. The army's elite combat and technology units not only provide training—often making up for what the schools fail to provide—but also form the basis of social and business networks that last long after recruits have completed their army service. Graduates of the most elite units, such as the vaunted 8200 intelligence unit and the air force pilots course, enjoy a pedigree in Israeli society equivalent to graduating from a top-flight academic institution in other countries.

Despite the dynamism ordinarily associated with Israel's start-up economy, high levels of income inequality aren't offset by high levels of wage mobility. A study looking at wage mobility for the years 2003–2009 found that 92% of Israelis in the top three deciles remained in the same decile from year to the next, a much higher rate than the 80.4% for European Union countries.[11] In the bottom decile, mobility was about the same—51.4% for Israel versus 53% for Europe—although income mobility for Israelis in this group declined considerably over the years.[12] In the middle second to fourth deciles, mobility was significantly higher—just 32.2% of Israelis were in the same decile from year to year, compared with 58.3% for Europeans. Compared to the USA and measuring changes over the course of the 1999–2009 decade, Israel also had lower rates of mobility. Looking at the highest fifth of wage earners, nearly two-thirds were in the same quintile at the end of the decade as they were at the start, a figure five percentage points higher than in the USA.[13] At the very top of the wage ladder, those earning the top 5% of incomes, Israeli mobility was higher than in the USA, but for the wrong reason: Those who had risen up to the highest wage brackets were disproportionately employed by companies operating in uncompetitive markets of one kind or another and were sharing in the fruits of monopoly profits.[14] In all events, after 2005 wage immobility grew steadily weaker in Israel, even though it was a period characterized by strong economic growth.[15] Moreover, the lower an Israeli stood on the wage ladder, the probability of his or her rising was smaller.

The phenomenon of growing inequality in Israel and the developed world is easier to document than it is to explain. Modern economies based on industry and mass consumer markets came hand in hand with increasing income equality over most of the twentieth century via virtuous circle of rising incomes and growing consumer spending. The assumption was that post-industrial economies based on knowledge would perpetuate the phenomenon, but that is not what has happened. "Widening income inequality is the defining challenge of our time," the IMF said in a 2015 report on the phenomenon.[16] Along with Israel, rising levels of inequality began to emerge in the Anglo-Saxon countries in the late 1970s, with others joining the trend in the late 1980s.[17] The pace accelerated with the onset the global recession in 2008, so that by 2011, the average income of the richest 10% of the population in OECD countries was about 9.6 times that of the poorest 10%, compared with seven times in the 1980s.[18] The Gini coefficient stood on average at 0.29 in the mid-1980s across OECD countries; by 2011/2012, it had increased to 0.32. Gini increased in 17 out of the 22 OECD countries for which there were adequate data available, rising by more than five points in Finland, Israel, New Zealand, Sweden, and the USA as well as Israel.[19]

Globalization and technological change have both been cited as a major factor behind the trend. The mass movement of manufacturing to emerging economies combined with automation taking the place of human labor in factory jobs has led to shrinking demand in a sector that had been a principal source of well-paying jobs for less-skilled labor in developed countries. While lower-skilled industrial jobs have disappeared, often replaced with lower-paying, less-secure jobs in the service economy, those with the greatest skills and education have been largely unaffected. Technological change has served to sharpen the process: Business investment has become focused on areas where productivity gains, hence wage increases, have accrued to those with the most skills and education. Thus, while education levels in developed countries have been rising, those without adequate training and schooling for the needs of the job market have been left further behind. A company that may have once encompassed a wide range of job skills from employees performing relatively low-skilled jobs on the factory floor, rising up to more skilled clerical jobs and then to those involving marketing and accounting, and finally to management and research and development has over the last two decades dispensed with the lowest categories. Manufacturing has been outsourced overseas and/or lost to automation and robotics, and clerical jobs have been replaced by computers or in some cases outsourced as well. What is left are the most knowledge-intensive roles. The OECD found that two critical aspects of globalization—trade integration and financial openness—had no significant impact on wage inequality or employment, but it did find that foreign direct investment (moving manufacturing to other countries, in this context) and increased financial flows did have an effect.[20]

In this context, the role of the high-technology industry (as against technology employed in business generally) in exacerbating inequality is simple enough to discern by its nature. Manufacturing is outsourced to low-cost countries while knowledge-intensive activities, like R&D, design, marketing, and management, remain in the company's home country. A study on wage and income inequality in the USA by metropolitan areas confirmed the correlation between wage inequality and knowledge-based and high tech, finding that inequality was greatest in small metro areas with a significant high-tech presence, such as San Jose, California, and Chapel Hill, North Carolina.[21] On the other hand, income inequality was more highly correlated to race, poverty, and what is termed the unraveling of the social compact.[22] The difference is that wage inequality is driven by wages at the top of the wage scale, while income inequality is more influenced by the bottom of the income scale.[23] A host of other factors have played a role to a greater or lesser extent, such as more women joining the labor force, the growth of assortive mating (people of the same income levels marrying as more women work at higher-paying jobs), and the aging of the population, to name a few.

Regulatory liberalization and changing tax policies have been cited as a factor as well. Starting in the 1980s with a policy revolution led by US President Ronald Reagan and British Prime Minister Margaret Thatcher, governments reversed a postwar trend of growing government intervention and higher tax rates in favor of policies that encouraged free markets and competition as a means of delivering economic growth and productivity. Unionization and job-protection legislation declined. As a whole, these policies have had the positive effect on generating more employment, but they also have contributed to growing wage disparities as more and more of the low-skilled had no choice but to enter the workforce.[24] Faced with shrinking demand for higher-paid and often unionized employment, low-skilled workers have found themselves in poorly paid service jobs, with few benefits and little job security. In the same vein, income-transfer policies were revised. From the mid-1990s onward, tax policies began failing to act as an effective counterweight to growing inequality[25] while benefits and allowances were increasingly allocated on a less-progressive basis.[26]

All of these factors that acted on the developed world—globalization, technology change, smaller government, and the rise of the high technology—have to a greater or lesser degree been felt in Israel as well. But the impact has been more pronounced, especially over the past 15 years, even if the most recent data show some improvement. In 2000, Israel's Gini index stood at 0.352 and rose steadily over the next six years to as high as 0.390 before leveling off and starting to show a decline after 2009 to 0.378 in 2011.[27] Israel was not alone is seeing inequality grow worse over the 2000s, but the deterioration started earlier and was more severe in Israel than in other OECD countries. In 1997, Israel was more unequal than the OECD average, but the gap was relatively narrow at 0.336 for Israel versus 0.299 for the organization's members. In the following years, the OECD Gini coefficient climbed but in 2007 was still just 0.303 while Israel's had risen to 0.369.[28] By 2012, as noted earlier, Israel's Gini coefficient had climbed to 0.371, the fourth highest in the OECD.

The trend toward smaller government through control of industries, rules and regulations, and income transfers was a relatively gradual process in Israel, but its impact was immense. As noted in Chapter 2, the late 1980s and early 1990s saw the rapid end to government and labor union dominance over the economy through a program of privatization and regulatory liberalization. Liberalization and privation made it easier for new and innovative businesses to develop, including Israel's start-up sector, enhancing economic growth and job creation. But the process failed to prevent—and arguably even fostered—the creation of private-sector cartels and powerful holding groups. Worse still, the liberalization process stalled, leaving key sectors of the economy as government monopolies. The result was a bifurcated labor market in which most workers were left without union representation and the social protections of the old economy while a narrower segment enjoyed generous pay and benefits because they were employed in state- or private-sector cartels. Meanwhile, much of the job growth in Israel in recent years has been generated by inefficient and labor-intensive sectors where jobs pay relatively poorly, a phenomenon discussed later in this chapter. High tech may have in a small way magnified the problem of inequality. The Israeli industry's narrow focus on start-ups engaged solely in innovation and engineering, as against manufacturing and marketing, means that the wealth it generates is shared by a relatively small number of people. Israeli high-tech companies provide very few jobs in the bottom income deciles, a moderate number in the middle deciles, and many more in the ninth and 10th deciles.[29] Moreover, high tech is a young industry and the high level of earnings companies pay typically lasts for a very limited number of years before employees are "aged out" and move on to lower-paying jobs in other sectors.[30]

After 2002, a second wave of the Israeli socioeconomic transition to smaller government and a greater reliance on free markets began as the government reduced allowances and restructured the tax system to favor earners over consumers. That had the positive effect of driving more Israelis into the labor market but reduced the incomes of those with the lowest pay. The share of Israeli households with no income other than government allowances dropped sharply from a peak of 13% in 2002 to just 8.8% nine years later,[31] while the number of breadwinners per household rose from an average of 1.17 to 1.31.[32] The problem was that many of these new entrants to the job market brought few skills and little education to their jobs and helped increase the ranks of the working poor.[33] The government might have acted to offset this—and did in a few instances—but the thrust of policy was to pare back its role in mitigating the effects of economic liberalization: The rate of civilian spending by the Israeli government relative to GDP fell about three percentage points in the decade to 2010 to just over 30%, the lowest rate among OECD members except for South Korea.[34] Although Israel spends close to the average OECD rate for health and education, in terms of direct aid to the lowest income groups Israel has become a very frugal spender, with only 10.9% of GDP in 2011 going to transfer payments of various kinds versus 17% on average for the OECD.[35] On the payment side of the ledger, Israel's tax regime evolved into a system that puts an unusually large burden on consumption spending through indirect levies like the value-added tax and a relatively light one on income. Thus, while Israel's income tax regime is highly progressive, a host of consumption-based taxes mean the overall burden is distributed in such a way as to exacerbate income inequality: In 2012, the bottom 20% of households by income had a tax burden of 30% while the top 20% paid just a little more, or 31%.[36] The middle three quintiles paid the least tax relative to their incomes, but even in that narrower group, higher-income households paid only fractionally more than their lower-income peers. As a result, the government's contribution to reducing inequality dropped by 10 percentage points in 10 years 25% in 2010, 10 percentage points below the OECD average.[37] Indeed, the Israeli government's role in mitigating inequality is so perverse that after removing the state's contribution and counting only earnings generated by households themselves Israel's Gini coefficient is relatively good. For 2011, Israel scored 0.494, so that among OECD countries eight had higher levels of inequality.[38] In other words, the economy itself was doing a relatively good job at distributing income equally; Israel's problem was mainly a failure of government policy.

In the classic trade-off between economic growth and income equality, Israel made the latter choice. Could Israel have done otherwise? Probably not. By the end of the 1990s and early 2000s, it was evident that the welfare burden was becoming too big for the government to bear and was acting as a deterrent for many to enter the labor market, a fact evidenced by the fact that Israel's labor force participation rate had fallen to dangerously low levels. Policymakers needed to create incentives to bring more Israelis into the labor market—and not simply a carrot policy of better training and education, but the stick of making it more difficult to remain outside the labor force. Reducing government allowances and evolving a tax structure that taxed income relatively less and consumption relatively more were both critical components. The sticks were an important tool vis-a-vis Israel's two traditionalist communities because the barriers preventing many from working are not simple economic considerations but bound up in social attitudes as discussed in Chapters 10 and 11. Haredi men have traditionally shunned work in favor of religious study well into adulthood while Arab women are discouraged from taking jobs outside the home or, when they do work, outside the immediate community. The government's task wasn't simply to create effective economic incentives to work, but incentives so powerful that they would gradually overcome social and religious mores. Although it has come at a high social—and economic cost—the alternative was worse.

The income and labor trends over the last decade have affected Israel's income groups in different ways—in short, the rich have grown richer, the poor have improved their relative position, and the middle class has struggled.

Since the outbreak of the social-justice protests in 2011, Israel's wealthiest have become the object of public scorn, in particular those whose fortunes derive from the cartels and holding groups that dominate the domestic economy (known in local parlance as tycoons) as well as the managers who serve them. In fact, there is some justice in the criticism because among the top 5% of wage earners in Israel in 2009—a group that counted just 4300 people, whose average salary rose 268% in the decade before—the biggest proportion (26.5%) worked in finance, an industry that is highly concentrated among a small number of banks and financial service companies.[39] Another group disproportionately represented is car importers, similarly cartelized industry. The country's high-tech elite have been mostly spared the same criticism, presumably due to the fact that their money is earned by virtue of their personal entrepreneurial abilities and doesn't come at any direct cost to consumers. Still, high tech was overrepresented among the wealthiest Israelis, accounting for 12% of the top 5% wage earners.[40]

Like other developed economies, inequality in Israel has worsened the most at the top end of the income scale, i.e., the wealthy have increased their incomes relative to the middle class and the poor. But relative to other developed economies Israel's wealthiest do not take a particularly high share of total income. The top 10% of Israeli income earners accounted for 24.9% of total income in the mid-2000s after taking into account taxes and government allowances, the sixth highest among 22 countries surveyed.[41] But that relatively high share was due to the tax policies Israel adopted over the last decade rather than due to economic distortions: Excluding their impact, the share of total income among Israel's top 10% was 30.3%, putting it only at No. 10 among the 22 countries. Where the obvious lack of conspicuous consumption emerges is Israel's relatively small proportion of income captured by people in the top 1%. That super-elite of income earners accounted for 5.3% of all income after taxes and allowances, making Israel eighth among the 22 countries.[42] Stripping out taxes and allowances, the rate was 6.3%, putting Israel at just No. 15. The data suggest that the monopoly profits that domestically oriented Israeli business generates are captured to a large extent by unionized employees at the expense of senior management and shareholders and thus filter down to what could be defined as the upper-middle class.

At the bottom of the economic ladder, Israeli poverty is unusually widespread relative to other developed economies. At 18% of the population in 2013, Israel's rate was higher than any other developed country belonging to the OECD and 64% higher than the average for the organization.[43] Child poverty, which is especially damaging because it more often than not reduces the chances of the upcoming generation escaping poverty in adulthood, was 23.5% in Israel in 2013, 74% higher than the OECD average of 13.5%. But that headline number should be kept in perspective. Israel has made significant strides in reducing its poverty rate over the last decade as the government adopted policies forcing people into the labor market. Among individuals, the rate climbed in the early years of the new policy, from 19.5% in 1999 just before as the policies began to be implemented to as much as 25% in 2009, but it has since retreated and was 21.8% in 2013, according to Israel's National Insurance Institute.[44] Among children, the rate of decline was even more dramatic: After rising from 26% in 1999 to a peak of 36.3% a decade later, the rate had fallen to 30.8% in 2013. Moreover, the depth of poverty in Israel is not particularly high by OECD standards: The average income for a poor family is 32% below the poverty line, which is higher than the organization average of 27% but still puts Israel just seventh from the bottom of the rankings.[45] Moreover, poverty is concentrated in two conservative, traditional-bound sectors where the influence of family, religion, and community mitigates the kind of social pathologies typically seen among the poor. The poverty rate among Israeli Jews was just 14%, which is still relatively high on the OECD ladder, but for ultra-Orthodox Jews it was 53% and for Israeli Arabs 54%.[46]

Overall, Israel's policy of coaxing more Israelis into the labor market while reducing financial assistance seems to have had a far bigger impact on reducing poverty than inequality because the economy favored job growth at the lower end of the education and income ladder more than at the top. Among the chief categories of Israelis who usually belong in the lowest income brackets—single-parent families, families with five or more children, Arabs, the ultra-Orthodox, and those with less than eight years of schooling—all saw their rate of labor force participation rise sharply in the decade after 2002, in most cases at a far faster pace than for the overall population.[47] In addition, wages for the lowest income groups have also risen, although it took longer for that to happen. In the four years prior to the global recession that started in 2008, job growth was led by Israel's high-tech sector and wage increases during that time were five times as high in job categories typically belonging to middle- and upper-income groups than those held by the poor.[48] But the trend changed after 2009, when salaries for high-income workers fell as the global recession hit industries like high technology but had less effect on lower-wage industries that cater to the domestic market. From 2010 to 2013, wages recovered, even for those paying the lowest wages even as the poor continued to enter the labor market.

It seems that business responded to the rising labor supply by creating low-paid, low-productivity jobs rather than investing in machinery and equipment or research and development, and the question is why. The answer may be because the social quality of the new entrants to the labor market. Despite all their disadvantages in terms of education, discrimination, and alienation from mainstream Israeli society, ultra-Orthodox Jews and Arabs live in communities with strong family-oriented values and a level of social stability that makes it easier for people to enter and adjust to the demands of the labor market than is typical for the poor and long-term unemployed. This is especially the case in the ultra-Orthodox world, where studying religious texts has a higher social value for adult men than holding a job, and unemployment is usually a matter of choice. Thus, when the financial pressures created by the government's policies became irresistible, the transition to employment wasn't difficult for either group. However, where the back-to-work policy has failed, it has failed in a big way: The poverty rate among Israeli families with no breadwinner reached 72.9% in 2013,[49] which is no surprise given that the government's contribution to reducing the rate of poverty was just 25.2% in 2012, compared with 36.4% a decade earlier.[50] Those who couldn't or wouldn't enter the workforce paid a heavy price.

The most important socioeconomic group for any developed economy is the middle class. Defining what constitutes the middle class is subject to an unresolved debate, but by almost every economic measure the middle class in the developed world is in decline even as economic life is more and more reliant on the skills, education, social stability, and aspirations the middle class provides.[51] Not just measured in income, the middle class is under pressure by rising living costs, most particularly for the three critical facets of middle-class life—housing, health, and education. Israel is no exception to the trend: Its middle class has declined in terms of size and relative income over the last decade. A study by the Bank of Israel, which defined the middle class as households earning anywhere between 75% and 200% of median income, found that the middle class' share of the population had fallen to 50.2% in 2011 from 51.6% in 2002.[52] Another measure of the middle class that sets a lower-income level of 50–150% of median income found a similar trend, with the middle class' share of the population shrinking from 55.1% in 2002 and 53.3% in 2011.[53] A third that examined a narrower range of incomes between 75% and 125% of the median found a drop from 28.1% in 2002 to 27.8% eight years later.[54] A fourth measure employed by the Taub Center for Social Policy Studies in Israel that divides the middle class into an upper- and lower-middle class found a slight but statistically insignificant increase in the total middle class's share of the population to 69.8% in 2011 from 69.3% in 2002.[55] But, Taub found the growth occurred only in the upper-middle class while the lower-middle class shrunk in those years.

All in all, the numbers don't point to a dramatic decline in the middle class, but it should be noted that vis-a-vis all three measures that in the base year 2002 Israel was in a deep recession and that the decade that followed were years of almost uninterrupted growth, yet the middle class' share of the population was lower or virtually unchanged from the 2011 figure. In other words, the decline of the middle class in Israel came against the background of a growing economy and rising levels of education and white-collar employment. Perhaps then it was no coincidence that the final year of the three surveys was the same year that middle-class Israelis took to the streets in mass protests against declining living standards and a sense that society was growing more unfair and unequal. It also marked the rise of "social" parties whose platforms have focused on the socioeconomic concerns of middle-class Israelis instead of the defense issues that have traditionally dominated the Israeli political debate. Still, the decline doesn't seem dramatic enough to have spurred such a strong reaction from ordinary Israelis. Rather, middle-class distress seems more a function of costs than income.

In Israel, that distress begins with the fact that the cost of living is higher than in most of the developed world, so that the challenge of maintaining a Western standard of living is not limited to the lowest income groups but to the middle class itself. That challenge has become more difficult over the last decade. A Knesset study found that consumer prices in Israel rose 34.1% between 2005 and 2013 while wages on average rose 2.3% after inflation.[56] The result was that prices, which in 2005 had been under the OECD average, were 10% higher by 2013 and 13.5% higher than they should have been.[57] For housing, a key expense item for middle-class families, Israelis were paying much more than their European counterparts do, with the price for a small 70-square meter unit costing the equivalent of more than 13 annual salaries in 2014 versus as little as 3.2 for Belgium and four to eight for most other European countries.[58] The expenses issue becomes more critical given that the middle class encompasses such a wide range of incomes, belying the conventional view of it is a unitary population. The middle class might have a lot in common in terms of shared values and aspirations, but in terms of material life the differences are considerable. A Knesset study of middle-class incomes, which it defined at households between the middle of the fourth decile and the seventh decline, ranged between 7600 shekels and 18,600 a month, a difference of 2.5 fold.[59] Those at the bottom rungs of the middle class are contending with a much more difficult financial situation than those at the top.

Looking at trends for middle-class households in the years before the 2011 social-justice protests, the Bank of Israel found that income growth exceeded expenditure growth for key goods and services like electricity and water, food, health, and preschool education for the decade to 2007, but that the trend reversed in the following three years.[60] The growing burden was especially heavy on younger families, who have lower incomes on average to begin with and generally have higher costs for housing (either because they are renting or are repaying bigger mortgages) and education (especially for preschool children). There are no more recent figures on income/expenditure trends, but one key item has certainly risen very sharply since 2010, namely home prices. Buying a home cost on average the equivalent of 146 monthly salaries in the first quarter of 2015, compared with just 43 months in 2008.[61] Under the circumstances, it is not surprising that using the government's 2011 household expenditure survey, the Taub Center found that Israeli families in all but the top 20% of income earners spent more in an average month than they earned, in most cases by considerable amounts. For the second highest quintile of households, the gap amounted to 5% of income and for the middle quintile to 10.7%.[62] For the lower two quintiles, of course, the shortfalls were considerably bigger. The Taub report speculates that the shortfall is closed by income unreported to the tax authorities, but concedes that the gaps are too wide to assume that the difference is made up solely by that source. Instead, it suggests that parental savings are routinely supplementing even middle-class families' expenditures.[63] In fact, a 2013 survey bears out this dangerous trend, finding that 87% of parents provided regular financial help to their adult children.[64] The assistance was considerable, with two-thirds of those surveyed saying they provided an average (in shekel terms) of just over $610 a month—an amount that comes close to the excess of expenses to income that the Central Bureau of Statistics found in its household survey. The phenomenon encompassed all income groups, not just lower-income families, and, needless to say, is not a sustainable model for family finance for more than a single generation.

In spite of high levels of inequality and the relative lack of wage mobility in Israel, other trends do suggest that Israeli society is capable of providing opportunities for social and economic advancement to its mélange of ethnic and religious groups as well as to women. The special cases of the Haredim and Israeli Arabs are dealt with in separate chapters, but vis-a-vis women, Russian and Ethiopian immigrants, and most notably Mizrahi Jews, Israel has made significant and usually overlooked progress.

Like their peers in the developed world, Israeli women have entered the workforce in increasing numbers over the last decades while improving their educational levels to the point where they are now obtaining tertiary degrees at higher rates than men.[65] In the 2013–2014 school year, women accounted for 57.3% of all students pursuing a degree at all levels through doctorate.[66] As more women enjoyed higher levels of education, their labor force participation rate rose between 1979 and 2010 from 44% to 70% (a faster pace of growth than in many other developed economies) at a time when the rate for men declined from 91% to 82%.[67] Increasingly higher rates of education have not only led to more women working—educated women at least—they have helped close the wage gap with men.[68] If women haven't succeeded in matching men's wages, it is due to the fact that more women work part-time (about a third of all Israeli women versus just 10% of men) and tend to work in professions that demand a post-secondary degree but pay relatively poorly. Women held 57.8% of all jobs requiring higher education, but they held less than a third of all managerial positions and they accounted for just 35.5% of those working in high tech,[69] which in the latter case isn't surprising since in fields like science and math, computer science, and engineering men still comprise the large majority of those pursuing degrees.[70] Women still accounted for two-thirds of all those working in traditionally "female" job categories like school teacher, bookkeeper, and retail sales, which traditionally pay poorly.[71] The degree to which women tend to lower-paying professions and are underrepresented in management roles is due to choice or to discrimination isn't an easy matter to determine in Israel or elsewhere. In any case, the progress women have made is incontestable and their achievements in education make it almost certain they will continue.

Russian and Ethiopian immigrants constitute the two biggest and most distinct immigrant populations to have arrived in Israel over the past 25 years, but apart from their newness to Israeli society and the challenges they have faced the challenges of integrating into a new and unfamiliar environment, they have little common. Both groups have suffered discrimination of the kind that immigrants inevitably experience in a new country, even one like Israel where immigration is regarded not just as economically beneficial but an essential part of the Zionist enterprise. Russians have had the advantage of a European background, although the fact that they immigrated in such big numbers in such a short period created economic anxieties and short-term problems in creating adequate and appropriate employment that the much smaller Ethiopian immigration did not. But, as noted in Chapter 7, Russian immigrants were highly educated and had a relatively easy time assimilating into the Israeli economy and society. One study of the development of Israeli middle class found that in 1992, when the great majority of Russians had just arrived in Israel, 56.7% of Russian immigrants were in bottom stratum of wage earners, another 32.7% were in the middle, and just 10.6% in the top.[72] By 2010, the percentage of Russians in the middle had remained about the same at 34%, but the number in the bottom stratum was now just 38.7%, an 18-point decline, while those at the top had nearly tripled to 27.2%. Russians were still overrepresented in the bottom stratum and underrepresented at the top relative to other Israelis Jews; nevertheless, it was a significant and rapid improvement. The second generation of Russian immigrants have more fully assimilated and almost certainly show better numbers.

By contrast, Ethiopians came from a poor, agrarian African economy so that on top of contending with deeper racial prejudice than Russian immigrants, they lacked the skills and training to easily integrate into Israeli society. In 2013, their household income was about a third less than the average nationwide.[73] Nevertheless, they have made impressive gains, albeit at a slower pace than Russian immigrants. Employment rates for the key age group of people aged 25–54 have risen from 50% in 2000 to 72% in 2009–2011, although they remain lower than the 79% for the rest of Israel's Jewish population.[74] In contrast Russian immigrants, Ethiopian progress is mainly evidenced among younger people, who unlike their parents have been able to better integrate into Israeli society and obtain education and training. Among Ethiopians aged 30–35 who arrived in Israel after age 12, only 5.7% held a tertiary degree in 2008, but among those who were educated in Israel the rate was 19.7%.[75] Even if that was less than half the 40.3% rate for non-Ethiopian Israeli Jews, the figure marks significant progress and has presumably improved since 2008. The occupational profiles of Ethiopians have also improved significantly between generations. Among Ethiopians aged 30–35 who were educated in Israel, some 21% were working in high-skilled jobs or management in 2008, less than the 39% rate for non-Ethiopian Israelis but more than twice the 9% for older Ethiopians.[76]

The status of Mizrahim—Jews who trace their origin to the countries of the Middle East and North Africa—has been the subject of much controversy for most of Israel's history. They arrived in large numbers starting in the early 1950s, mostly housed in transit camps and eventually in isolated "development" towns where access to education and jobs was severely limited. Over the next 20 years, Israel's Mizrahim evolved into an underclass. In 1977, they staged a political revolt against Israel's

Ashkenazi establishment by giving Menachem Begin's Likud Party enough votes in the elections that year to end nearly three decades of uninterrupted Labor Party rule. The direct impact of the Likud victory was to put Mizrahi political leaders into positions of power, but more importantly it broke the lockhold of the Ashkenazi Labor establishment on the economy.[77] Against that, globalization and economic liberalization in the following years disproportionately hurt Mizrahim working at low-wage jobs in industries unable to compete in the changing economy. But already by the mid-1980s Mizrahim were starting to improve their financial and social status mainly by starting small businesses; the next generation advanced further by taking advantage of the rapid expansion of the higher education system over the following decade.[78] Intermarriage rates between Mizrahim and Ashkenazim have grown, initially as higher-income Mizrahim married low-income Ashkenazim but gradually evolving into unions of equals.[79] By 2011, the incomes of second-generation Mizrahim had reached 73% of their Ashkenazi peers, up from just 60% in the mid-1990s.[80] A wide income gap still existed, but it was closing at a rapid pace.

As is the case with women and Ethiopians, Mizrahim responded to the changes in the labor market favoring educated workers and pursued a higher education in increasing numbers.[81] While Israelis across the social and economic spectrum were getting tertiary degrees at increasingly higher rates, among Mizrahim the increase outpaced the rate of growth for Ashkenazim.[82] Education has enabled Mizrahim to be represented in the highest income groups proportionate to their share of Israel's population, and among middle-income groups, their gains have been considerable. At lower-income levels, less-educated Mizrahim have also lifted themselves out of the lowest income deciles, but that unfortunately was not due to rising incomes lifting them into higher deciles but because growing numbers of poor Haredim and Arabs have taken their place.[83] Ironically, Mizrahi social progress has been achieved by their mirroring the same social and economic cleavages that characterize broader Israeli society.

Like other developed economies, Israel faces not only the problem of growing inequality but also the more vexing problem of squaring the inevitably of the rise of a knowledge-based economy and globalization, and their obvious benefits in terms of rising incomes, greater economic efficiency, and rewarding labor, with the social and economic dislocations they have created. Far from encouraging equality, these two developments have begun reversing the gains made during the first two-thirds of the twentieth century. The answer would seem to be that if economic fundamentals can't address the problem, then it is the role of government to create tax and welfare policies that do what the market has failed to. But probably the most important of all is education policy, i.e., ensuring that the great majority of the population has the means to acquire the knowledge and skills to work and play a meaningful, productive role in the economy. Israel has had to cope with this problem earlier and more deeply than most of the developed world, giving it more time and a greater incentive to address the problem. But to date it hasn't done so in a fundamental way even if policymakers are aware of the extent of the problem.

Notes

1. Federico Cingano, *Trends in Income Inequality and Its Impact on Economic Growth, Organization for Economic Cooperation and Development* (Paris, December 9, 2014), 18. The report doesn't relate to Israeli growth, but estimates that growing inequality reduced by more than 10 percentage points economic growth in Mexico and New Zealand in the years 1990–2011. In the USA, Britain, Sweden, Finland, and Norway, the growth rate would have been more than one-fifth while greater equality helped increase per capita GDP in Spain, France, and Ireland. Other observations were taken from Era Dabla-Norris, Kalpana Kochhar, Nujin Suphaphiphat, Frantisek Ricka, and Evridiki Tsounta, *Causes and Consequences of Income Inequality: A Global* Perspective (Washington, June 2015).
2. Organization for Economic Cooperation and Development (2015), Income Inequality (database).
3. Ibid.
4. Ofer Cornfeld and Oren Danieli, "The Origins of Income Inequality in Israel—Trends and Policy," *Israel Economic Review*, Vol. 12, No. 2 (2015), 51–95, 76.
5. OECD, *Education at Glance 2014: OECD Indicators* (Paris, October 2014), 194.
6. Israel Education Ministry, *Meitzav 5775 Part A, Achievement Tests* (Hebrew) (Jerusalem, November 2015), 35, 36 and 43.
7. Bank of Israel, *Annual Report 2013* (Jerusalem, April 2, 2014), 209–210.
8. Yuval Dror, *Israelis in the Digital Age, 2012* (Hebrew) (Rishon LeZion Israel, May 2012), 12.
9. Maor Milgrom, *Economic Segregation in Israel, 1983–2008* (Hebrew) (Tel Aviv, February 2015), 11. The survey examined Israel's four

metropolitan areas and used an index of 0 for complete equality to 2 for full segregation. Israel's segregation index rose from an average of 0.26 in 1983 to 0.43 in 2008, an increase of 65%. In the USA, the index rose from 0.32 to 0.46 in the same years.

10. Inbal Orpaz, "The Secret to High-tech Success? This Elite Israeli Army Unit," *Ha'aretz*, April 18, 2014. The story didn't indicate for what period the figures covered.
11. Galit Ben-Naim and Alexey Belinsky, *Wage Divergence in Israel: Analysis of Wage Mobility Over the Last Decade* (Hebrew) (Jerusalem, June 11, 2013), 4.
12. Ibid., 5.
13. Ibid., 6.
14. Ibid., 7.
15. Ibid., 9.
16. Dabla-Norris et al., *Causes and Consequences of Income Inequality*, 4.
17. OECD, *Divided We Stand: Why Inequality Keeps Rising* (Paris, December 2011), 22.
18. Cingano, *Trends in Income Inequality*, 8 and 36.
19. Ibid., 8.
20. OECD, *Divided We Stand*, 29.
21. Richard Florida and Charlotta Mellander, "The Geography of Inequality: Difference and Determinants of Wage and Income Inequality Across US Metros," *Regional Studies*, 2014, 6.
22. Ibid., 8.
23. Ibid.
24. Ibid., 30–31.
25. Ibid., 37.
26. Ibid., 38.
27. Cornfeld and Danieli, *Origins*, 80.
28. Ibid., 59.
29. Ben-Naim and Belinsky, *Wage Divergence in Israel*, 11.
30. Meirav Arlosoroff, "Israel's Fat High-tech Salaries Are Suddenly Going to Look a Lot Thinner," *Ha'aretz*, November 13, 2015.
31. Cornfeld and Danieli, *Origins*, 81.
32. Ibid., 82.
33. *Report of the War on Poverty Committee, Part I* (Hebrew) (Jerusalem, June 23, 2014), 53. The rate of working poor among those below Israel's poverty line rose to 56% in 2012 from 38% 10 years earlier.
34. Welfare Policy Issues, Chapter 8, Bank of Israel 2013 Annual Report, 202.
35. Ibid., 203.

36. Ibid., 206.
37. Ibid., 202.
38. OECD (2015), Income Inequality (indicator).
39. Ben-Naim and Belinsky, *Wage Divergence in Israel*, 7. The top 5% are employed by 70 companies, whose 80,000 employees saw their average salary decline in 2003–2009 of 7%.
40. Ibid., 7.
41. Dan Ben-David and Haim Bleikh, "Poverty and Inequality Over Time In Israel and the OECD," *State of the Nation Report 2013* (Jerusalem, November 2013), 41–42.
42. Ibid., 43.
43. National Insurance Institute, *Dimensions of Poverty and Social Gaps, 2013* (Hebrew) (Jerusalem, November 2014), 42. The OECD defines poverty as 50% of a country's median income, so that the rate is not absolute across the OECD. Unlike Israel, it adjusts the rate according to the size of the family, so that its figures differ from those reported by Israel's National Insurance Institute.
44. Ibid., 20.
45. *War on Poverty*, 51.
46. Ibid., 52.
47. NII, *Dimensions of Poverty and Social Gaps, 2013*, 14.
48. Ibid., 13.
49. Ibid., 7.
50. *War on Poverty*, 50.
51. Rakesh Kochhar, *Middle Class Fortunes in Western Europe* (Washington: Pew Research Center, April 24, 2017), 5. The report found that the middle-class share of the adult population fell in seven of the 11 Western European countries examined in 1991–2010, mirroring the long-term shrinking of the middle class in the USA.
52. Israel, Bank of, "The Middle Class in Israel," in *Annual Report 2011* (Jerusalem, March 28, 2012), 343; and Zoya Nisanov, *The Middle Class in Israel* (Jerusalem, July 2014), 10.
53. Nisanov, *The Middle Class in Israel*, 10.
54. Noga Dagan-Buzaglo and Etty Konor-Attias, *Israel's Middle Class 1992–2010: Who Are We Talking About* (Tel Aviv, January 2013), 9.
55. Nisanov, *The Middle Class in Israel*, 13. Nisanov employs a mixture model developed by McLahlan and Peel.
56. Knesset Research and Information Center, *Data on the Cost of Living in Israel in Comparison to Developed Countries: An Update* (Itamar Milrad) (Hebrew) (Jerusalem, November 12, 2014), 1.

57. Ibid., 2. The study cited a correlation between per capita GDP and dollar prices and on that basis estimates that Israeli consumer prices should have been 96.9 on an index (OECD average = 100) but were in fact 110.
58. Deloitte Touche Tohmatsu, *Property Index Overview of European Residential Markets*, 21.
59. *The Weighting of the Middle Class and an Analysis of Changes in Recent Years* (Ilanit Bar) (Hebrew) (Jerusalem, April 14, 2013), 3.
60. Israel, Bank of, "The Middle Class in Israel" *Annual Report 2011*, 348.
61. Nimrod Bousso and Zvi Zrahiya, "Buying a Home Will Cost an Average Israeli 146 Monthly Salaries," *Ha'aretz*, September 30, 2015.
62. Eitan Regev, "Making Ends Meet—Household Income Expenditures and Saving in Israel," *State of the Nation Report 2014* (Jerusalem, June 2013), 65.
63. Ibid., 64.
64. Keren Tsuriel Harari, "The Middle Class' Secret Oxygen Supply: We Have an Account with the Parents," *Calcalist*, June 27, 2013. The poll of 800 people conducted by the polling organizations Midgam and iPanels excluded soldiers and adult students.
65. Ayal Kimhi, "Trends in the Labor Market: Gaps in Employment Rates and Wages," *State of the Nation Report 2011–2012* (Jerusalem, undated), 124. Among men, the percentage with 12 or more years of education was 52% in 2010, up from 42% in 1998, while for women the rate was 62%, rising from 52%.
66. Central Bureau of Statistics, "International Women's Day 2015," 9.
67. Kimhi "Trends in the Labor Market," 108.
68. Ibid., 123.
69. Central Bureau of Statistics, "International Women's Day 2015," 4.
70. Ibid., 9.
71. Ibid., 4.
72. Dagan-Buzaglo and Konor-Attias, 15. The middle stratum is defined as between 75 and 125% on Israel's median income.
73. Hadas Fuchs and Gilad Brand, *Education and Employment Trends Among Ethiopian Israelis* (Jerusalem, June 2015), 1. Ethiopians constituted only 1.7% of Israel's population in 2013.
74. Ibid., 4.
75. Ibid., 2.
76. Ibid., 11.
77. Nissim Leon, "Moshe Kahlon and the Politics of the Mizrahi Middle Class" (College Park, MD, January 2015), 8.
78. Ibid., 5, 7.
79. Momi Dahan, *Has the Melting Pot Succeeded in the Economic Field?* (Hebrew) (Jerusalem, September 2013), 4.

80. Ibid., 15.
81. Ibid., 26.
82. Ibid., 21.
83. Ibid., 16, 17.

CHAPTER 10

Outsiders I—Israeli Arabs

For a country of just over eight million people, Israel has a remarkably large number of distinct populations. However, Israeli Arabs and ultra-Orthodox (Haredi) Jews stand out in this constellation both because they comprise a large and growing share of the country's population and because they largely exist apart from the rest of Israel, not only economically but socially and culturally as well. For the most part, they live in their own towns and neighborhoods, their children learn in separate schools, speak their own languages, and have their own community leaders and cultures. On an ideological plane, neither Israeli Arabs nor Haredi Jews sees themselves as part of the Zionist enterprise—whose acceptance has traditionally been the ticket of admission to be accepted as a full member of Israeli society. Neither Israeli Arabs nor Haredim as a rule serve in the army, which as noted in Chapter 8, plays a critical role in the formation of young Israelis' values and social networks to a degree perhaps unparalleled anywhere else in the world. Religion, or lack thereof, is a major element of personal identification for Israelis, the subject of endless opinion polls and political debates. In the Israeli context, it dictates how you dress, where you will eat, and who are your friends and neighbors, so the fact that Haredim practice a highly intense form of Judaism and Arabs are Muslim or Christians does a lot to set them apart. These separating factors all play a critical role in whether an Israeli sees himself or herself as being on the inside or the outside of what can be called mainstream Israel—the majority of Israelis who identify with

D. Rosenberg, *Israel's Technology Economy*, Middle East in Focus,
https://doi.org/10.1007/978-3-319-76654-6_10

the country's institutions, participate in its popular culture, and share the country's dominant values and social norms.

Over the first decades of Israel's existence, the apartness of Israeli Arabs and Haredim was largely ignored; arguably, it prevailed as a matter of mutual tacit consent. Economic and social isolation meant that Israeli Arabs and Haredim were worse off economically than the mainstream Jewish population, but that was the price to be paid for the cultural autonomy that both groups—especially ultra-Orthodox Jews—aspired to. For the rest of Israel, the outsider status of the two groups allowed a young society to develop its political, social, and cultural ethos without having to acknowledge the special needs and differences of the two minorities. In any case, the fact that Israeli Arabs were identified with the Arab and Palestinian enemy created an atmosphere of distrust that would have complicated any attempt at social and economic integration. Haredim were fellow Jews, but they were even more intent on cultural autonomy than Israeli Arabs and looked askance at mainstream Israeli society. For a long time, the economic cost of this apartness, which meant that the labor and talents of Israeli Arabs and Haredim were not being fully utilized, was tolerated. But as the two groups' share of the population has grown and the population of mainstream Israel ages, the situation has become unsustainable, a fact that Israeli policymakers have begun to acknowledge in the second decade of the twenty-first century.

In 2009, the population of mainstream Israel, comprising non-Haredi Jews and a tiny number of non-Jews, constituted 69.7% of the population, but projections by the Central Bureau of Statistics show their share declining to 59.7% in 2034 and to just over half in 2059.[1] Haredim, which in 2009, accounted for 9.9% of the population, are expected to see their share grow to 17.2% 25 years later and 26.6% a half a century later. Those are mid-range projections: The Haredi share could grow to as much as 21% in 2034 and 40% in 2059. Counting the ultra-Orthodox population, much less projecting its future growth, is a difficult business because definitions of what constitutes "Haredi" vary and given that it is a community defined by belief and practice, members drop out, and others join.[2] Nevertheless, even by the lowest CBS projection of 14% and 16% in 2034 and 2059, respectively, the ultra-Orthodox share will grow. The Israeli-Arab share will grow less dramatically, but it is already more than twice that of the ultra-Orthodox, making up 20.3% of the population in 2009. That proportion is projected to increase to 23% in 2034 and remain at the level at least to 2059. A third important demographic concerns the aging of Israel's population over the next decades. The CBS forecast estimates that the population over age 65 and no longer in the workforce will grow from 9.8% in 2009 to as much as 15% in 2034 and 21% in 2059.[3]

These trends raise two immediate concerns from the point of view of the economy. The first is that it spells a sharp decline in the size of the labor pool relative to the population and, as a corollary, an increasing number of dependents per workers. As is discussed in this chapter, the labor force participation rate for Israeli Arabs and Haredim is very low and especially low for Israeli-Arab women and Haredi men. Yet, according to the CBS projection, the share of the population for these two subgroups in the key working age population (ages 20–64) will grow from 12.4% of Israel's population in 2009 to 19% in 2034 and 24.8% in 2059.[4] The second concerns Israel's ability to retain the start-up ethos discussed in Chapter 8 unless Israeli Arabs and Haredim can be fully integrated into mainstream Israel, which is an issue discussed at the end of this chapter. Both groups are gradually moving into the workforce and/or enhancing their skills and education, but the process has been very slow and replete with political and social obstacles. The apartness phenomenon complicates the process by erecting barriers to assimilation into mainstream Israel and may ultimately make it impossible to fully realize.

From a socioeconomic perspective, Israeli Arabs and ultra-Orthodox Jews share much in common in terms of high levels of poverty exacerbated by large families, less formal education than other Israelis, low rates of labor force participation, and lower incomes. But the history and social factors that created this situation are so different for each group that they should be treated separately.

Israel's Arab minority are the indigenous Palestinians who remained in Israel after the 1948 war that created the state. The historical facts of the war, in particular the extent to which Palestinians left voluntarily to escape the fighting or were forced from their homes, are a matter of intense debate to this day because of its political implications. From the perspective of Israelis, the war was imposed on them after Palestinians and the Arab world rejected a United Nations-imposed division of the land into two states. As violent conflicts go, Israeli forces fought fairly and with restraint against overwhelming odds. From the Palestinian perspective, the Jewish presence in historic Palestine is unjust to begin with. They cite scholarship that has emerged since the 1980s documenting mass expulsions of Palestinian civilians and killings.[5] In all events, the war left the Palestinians inside the newly created State of Israel (or Israeli

Arabs, to distinguish them from the Palestinians outside the country) traumatized. Some 720,000 people, amounting to 85% of the Palestinian pre-war population, had left, turning them in the space of 18 months from a majority into a minority of just 18% of the new state's population.[6] That figure would rapidly decline over the next decades to as little as 11% as waves of immigrants arrived in Israel and would not reach its 1948 level again until 1985. Moreover, the Palestinians who left during the fighting included the great majority of the educated urban elite, leaving a population that was mostly rural and illiterate and without the institutional resources to overcome the loss.[7] During the era of the British Mandate, Palestinians hadn't created the kind of network of educational institutions the Jewish community had; Palestinians seeking a higher education went to Cairo, Beirut, or Europe. When the war ended, there was no institutional base for Israeli Arabs to build a new business, political, and intellectual leadership, and because that Arab world rejected any kind of contact with the Jewish state, pursuing a higher education in an Arabic-speaking country abroad was no longer an option. The War of Independence, as it is known in Israel, came to be known as the *Nakba* (catastrophe) among Palestinians.

The scope of this book is confined to the issues of human capital, but examining the issue of Israeli Arabs' place in Israeli society and their ability to integrate into mainstream Israeli society is inseparable from this history. Although Israeli law on the whole accords them equal rights and Arab parties have been represented in the Knesset since the first elections, Israeli-Arab towns were under military rule until 1966 and expressions of political or national identity were constrained. For nearly five decades, Israeli Arabs were characterized by political and national quiescence. That situation began to change in the 1990s and the onset of the Oslo peace process between Israel and the Palestine Liberation Organization, which spurred hopes among Israeli Arabs that the creation of a Palestinian state by side with Israel would normalize their status and at the same time renewed their sense of national identity.[8] The rapid unraveling of Oslo dashed those expectations. By the year 2000, when the last remnants of the process dissolved into the Second Intifada, unrest in Israeli-Arab towns led to clashes with police and 13 deaths, the process of a growing Palestinian consciousness and a concomitant distancing from Israel was well underway. One way of measuring these changing attitudes is an annual poll asking Israeli Arabs how they identify themselves. In 1995, while the Oslo process was still underway, 53.6%

of Israeli Arabs still identified themselves as Israeli Arabs; by 2012, the figure had fallen to 32.6%. By contrast, those defining themselves as a Palestinian Arab in Israel, Palestinian or Palestinian Arab had grown from 46.4% to 66.5%.[9] In other words, in 2012 more than two-thirds defined themselves as exclusively or mainly Palestinian Arab and just over 11% as exclusively/mainly Israeli (the rest described themselves as both equally).

The implications of this self-definition are not easy to discern. Politically, Israeli Arabs have stopped voting for mainstream (Zionist) parties in favor of Arab parties that to one degree or another stress their Palestinian identity and hostility toward Israel as a Jewish state.[10] But in the context of the practical choices Israeli Arabs face in their day-to-day lives and their interactions with the Jewish majority—all of which have much more bearing on issues relating to the economy and human capital—survey responses create a much more nuanced picture. Asked in a 2015 poll, for instance, whether they would be willing to see a Jewish doctor, 91.4% of Israeli Arabs said yes, a higher percentage than the 77.8% of Jews expressing a willingness to see an Arab doctor.[11] Some 84% of Israeli Arabs would accept a Jewish teacher for their children, while 74.4% of Jews say the same vis-a-vis an Arab teacher.

Numbers like these don't point to perfect social integration, but they should be viewed from the perspective that relations between Jews and Arabs in Israel are strongly influenced by the country's unresolved conflict with its Arab neighbors, so that for both sides the image of the other is to one degree or another the image of the enemy. In this context, it is not at all surprising that vis-a-vis issues of deep social integration views on both sides are less favorable. On intermarriage, 38.8% of Israeli Arabs and 36.8% of Israeli Jews actively oppose the idea to the degree that they expressed support for fringe organizations fighting it. In another poll from 2013, 29.7% of Israeli Arabs said they would be unwilling to have an Israeli Jew as a neighbor while close to 45.7% of Jews expressed the same feelings about having an Arab neighbor.[12] However, close to 85% of Israeli Arabs in a survey expressed a strong desire for their children to learn Hebrew at a young age.[13] Like the flag, the army or the collective memory of the Holocaust, Hebrew is an integral part of Israel's mainstream culture; for Israeli Arabs, Arabic is one of the most important elements of their separate identity. However, Hebrew fluency has economic benefits, enabling Israeli Arabs to more easily pursue a higher education and compete in the job market. The controversy of civilian national service for Israeli Arabs similarly points up the dilemma. Inside the community, service is controversial because of the program's links with the army and because many assert that Israel must first accord full rights to its Arab minority before it can impose obligations. Despite that, the number of Israeli Arabs participating as volunteers in schools, hospitals, and other institutions as an alternative to army service doubled between 2005–2006 and 2012–2013 to 3600.[14] Most of the volunteers coming from the middle class and are more likely to college-bound,[15] which suggests that volunteers look at it as a way of enhancing their educational and career prospects even as community leaders reject it for political reasons.

In short, Israeli Arabs face a dilemma: Most seek economic and even social integration into wider Israeli society but resist full identification with the state. On the surface, this doesn't seem like an impossible goal; democracies can and should be able to absorb different levels of political loyalty and national identification. Israeli is a polyglot society, but it makes considerable demands from its citizens and is characterized by a high degree of social cohesion, thus dual identity and affiliation with Israel for pragmatic purposes is seen by many in mainstream Israel insufficient, perhaps suspect. One reading of this is evidenced in a 2015 survey that found that while more than three quarters of Israeli Arabs accepted the view that "Arab citizen of Israel who considers himself an integral part of the Palestinian people [can] be a loyal citizen of the State of Israel," only a third of Israeli Jews did.[16]

Not just identity, more concrete factors inhibit social and economic integration. Israeli Arabs live for the most part geographically separated from Israeli Jews. Not counting East Jerusalem, 44% of the Israeli-Arab population is concentrated in 14 localities, most of them exclusively Arab cities like Nazareth and Um al-Fahm.[17] In mixed cities, like Haifa, Tel Aviv-Jaffa, Ramle, and Lod, Israeli Arabs live in separate neighborhoods, even if life involves more daily contact in the workplace, shopping and dealings with official bodies. Israeli-Arab towns suffer from far lower levels of public services, infrastructure and local employment, including a critical lack of public transportation and poor roads for those with private vehicles. As of 2009, for instance, 41% of 147 Arab localities had no public transportation at all, and another 43% had only limited services.[18] The problem of transportation is usually framed in terms as an obstacle to Israeli Arabs finding work, but it presents it also creates a barrier to integration by preventing Israeli Arabs from reaching shared places with Jews for activities like shopping, entertainment, and culture. Israeli

Arabs are less likely to use the Internet frequently if at all, compared with Israeli Jews, putting them at a disadvantage in terms of computer-related skills and connectedness to the wider Israeli world.[19]

The apartness of Israeli Arabs is not limited to the issue of identity but extends to their socioeconomic condition. Compared to non-Haredi Jewish Israelis, Israeli Arabs suffer much higher rates of poverty and much lower incomes; they are less likely to be employed and less likely to have a post-secondary education. In 2013, 52.4% of Israeli Arabs were under Israel's official poverty line even after taking into account the effect of taxes and transfer payments, a rate was two-and-half times the overall national rate and more than three-and-half times the rate for non-Haredi Jews.[20] As it has for Israelis generally, poverty among Israeli Arabs has risen over the last two decades, peaking in 2011–2012, although in more recent years, the increase has leveled off and has even shown some signs of declining again, as more and more people enter a workforce that has shown a surprising ability to absorb people at the lower end of the skills and education ladder.[21] Thus, the poverty rate for Israeli-Arab families dropped close to seven percentage points in 2013 alone before the impact of transfer payments.[22] Still, the impact of growing employment among Israeli Arabs would have to be sustained for quite a long time to bring the poverty levels back even to the level of 38.7% in 1997. The issue of poverty is exacerbated by high birthrates and the fact that Israeli Arabs live apart from mainstream Jewish Israel in their own towns and neighborhoods where poverty is the norm, and there is a paucity of social services and quality schools. This is all captured in a government index for socioeconomic conditions that ranks Israeli cities on a scale of one to 10, with 10 having the best parameters for education, employment, vehicle ownership, and standard of local infrastructure. In Israeli-Arab localities, the median rank is three and no city exceeds six.[23] In mixed cities, the average rank is four, and in Jewish localities, it is six. A dysfunctional culture of poverty may not exist among Israeli Arabs (about which more below), but in terms of government activity to ameliorate or reverse it, relatively few resources have been deployed.

The factors behind high rates of Israeli-Arab poverty are low levels of employment and income, and high rates of joblessness. But in examining the dimensions of the problem and many of the causes behind it, it is more useful to break down the labor force by sex rather than to look at it is as a whole. Among Israeli-Arab men, the labor force participation rate has been declining steadily over the decades from 75% in 1970 to about 63% in 2011 for those ages 15 or over.[24] The participation rate for men in developed economies, including Jewish men in Israel, also declined in those years, but not nearly as rapidly. More recent data show that the labor force participation rate for men and women in Israel has begun rising, spurred by a growing supply of low-skill low-pay jobs again as noted in Chapter 7,[25] but it seems unlikely the trend can sustain itself for long because Israel can only compete in the global economy in sectors that require high standards of training and education. In that respect, Israeli-Arab men are a poor match for the market: In the 18–45 age group, only 25% had more than 13 years of schooling in 2010, less than half the 52% rate for Jewish males.[26] It necessarily follows that Israeli-Arab men are heavily concentrated in low-skilled labor. In 2011, 12.9% of them were in unskilled jobs, nearly double the rate for Jewish males.[27] Only 15% held jobs that required a post-high school education of some kind, just over a third the rate for Jewish males. Their education and job profile has put Israeli Arab men in competition over the last decades with Palestinian laborers from the West Bank and Gaza Strip and from imported guest workers, both of whom are prepared to work for lower salaries in the lowest-skilled jobs. And, because these jobs often require considerable physical labor, Israeli Arab men tend to leave the workforce at a relatively young age—starting at age 45 versus age 60 for Jewish men—which goes a long way to explaining the low overall rate for labor force participation.[28]

Israeli-Arab men's disproportionate presence at the bottom end of the labor force has also meant that their wages are unusually low. At an average of 33 shekels an hour in 2011, they were not only 43% less than for Jewish men[29] but also slightly under the average of 35 shekels for Arab women[30]—a figure that serves as one indicator of the huge gender gap among Israeli Arabs. Far more than men, Israeli Arab women have exploited the growing opportunities over the last two decades to obtain a higher education and make use of their credentials in the job market. The percentage of Arab women with 13 years of schooling was 27% in 2010, an increase of nearly 3.5-fold in the space of two decades and a higher rate than among Arab males.[31] Better training and education—as well as fewer opportunities for women in low-skilled jobs—has meant that Israeli Arab women are far more likely to be in higher skill, better paying jobs. A fifth are employed in education and another 16% in health and social services; they are more likely than Arab men to work in business and finance and are even more likely to hold a job requiring a post-high school degree than Jewish women (44.7% versus 40.2%).[32]

Despite the relatively high employment profile, the wage gap with Israel Jewish women is wide, but at 21%, it is less than half that between males.[33]

The problem is that the job categories filled by Israeli-Arab women have become saturated (especially as Arab women in the public sector tend to work in their own communities) and even those with a tertiary education struggle to find work in the mainstream Israeli economy. Moreover, the relative success of Israeli-Arab women has to be measured against their extraordinarily small presence in the job market. Their labor force participation rate has more than doubled in the 20 years from 1990, but that was from a very low baseline of about 10%.[34] By 2011, it stood at 22%, less than a third the rate for Jewish women. The low rate suggests that the relative success of Israeli-Arab women may have a lot to do with self-selection, that is, only a minority of the best and brightest seek to enter the labor force at all, so that those who do are more likely to thrive.

One alternative for Israeli Arabs is to go into business, and in fact, what little data there are suggest that the rate of entrepreneurial activity for Israeli Arabs is somewhat higher than for Jews. However, the kind of entrepreneurship Israeli Arabs practice is on a small scale—family-owned businesses like restaurants and building-materials suppliers, serving the immediate community, with no strategies for innovation or growth. Some 80% of employ fewer than two workers, and management and staff are typically drawn from extended family or neighbors.[35] Certainly, government neglect of Israeli-Arab cities has contributed to this: Most don't have designated industrial zones that could serve as a home for bigger, more ambitious businesses, and few Israeli Arabs have the connections needed to navigate the process of winning government investment and research and development grants.[36] Israeli-Arab entrepreneurs face discrimination from banks for finance, so they are dependent on family capital, which in turns means they are often compelled to employ family members as, in effect, a return on the family's investment.[37] Given the Israeli start-up sector's apartness from the Israeli economy, none of these factors should be acting as a deterrent to Israeli Arabs forming high-tech companies. But the record has been poor, although there are several programs aimed at supporting start-ups for the community.

Over the last decades, more and more Israelis Arabs have pursued a higher education, mirroring the increase throughout the population.

But the rates for Israeli Arabs are still well behind the Jewish population, and their ability to match them is severely constrained by a variety of factors—poor student performance at the elementary and high school levels, high dropout rates, insufficient spending on education, and the language barrier they meet as young people growing up in a principally Arabic-speaking environment enter Hebrew-speaking colleges and universities.

Education through high school for the great majority of Israeli-Arab children takes place in separate schools, where the language of instruction is Arabic and the curriculum slightly different from what Jewish Israelis are taught,[38] which draws a sharp line both in terms of the resources available to them and educational outcomes. Classroom crowding in Israel is high in general relative to other developed economies, but in Israeli-Arab elementary schools it is higher still—27.4 pupils versus 26.8 nationwide.[39] Nearly a third of all Arab students are taught in non-standard facilities like rented buildings, their schools have fewer computers for student use,[40] and they get less money per pupil, although this in part reflects the inability or unwillingness of Israeli-Arab local authorities to supplement budgets from the Education Ministry.[41] The annual test of student achievement for fifth and eighth graders administered by the Education Ministry shows students in Hebrew-language schools (i.e., Jewish Israelis) outperforming their peers in Arabic-language schools by 21–82 points in the 2007/2008 to 2014/2015 school years across all grades and all subjects, with no evidence of the gap closing.[42] These gaps are confirmed by the Program for International Student Assessment (PISA) tests administered by the Organization for Economic Cooperation and Development at eighth grade, which show a wide and growing difference between Israeli Jewish and Israeli-Arab scores since 2000 of about 100 points.[43] Students from lower socioeconomic backgrounds, as Israeli Arabs overwhelmingly are, typically perform more poorly than those in higher groups regardless of their ethnicity. But the PISA scores for Israeli Arabs are consistently lower than for their Jewish peers even when they are from the same income group, suggesting that other factors are at work. In math, for instance, the gap between Israeli Jews and Arabs scores is 67 points for the lowest socioeconomic groups, widening to 95 for the middle group and to 106 points for the highest.[44]

Despite the inferior school performance, more Israeli-Arab high school graduates are continuing on to higher education, albeit at much lower levels than Israeli Jews. The proportion of Israeli Arabs ages 14–17 in school rose to 93% in 2014 from 79% 15 years earlier, narrowing the gap with the Jewish population to just four points from 16.3.[45] The percent who met minimum entrance requirements for university likewise climbed more than 16 points to 42% in 2014, although still far below the 52% for Jewish students.[46]

Unlike the primary and secondary schools, Israel's colleges and universities are fully integrated; indeed, they are often the place that young Israeli Arabs have their first day-to-day contact with the Jewish majority. However, the apartness of Israeli Arabs has already been firmly established in their childhood and adolescence. When they arrive on campus, they are separated from the Jewish majority not only by language but also by age, since Jewish Israelis typically start their higher education well after high school because of army service and other delays. They are more mature, have a better sense of their goals, and often bring critical skills and experience from the army. Entrance exams and instruction are in Hebrew, which puts Arab students at a disadvantage academically, and the emphasis on rote learning in Israeli-Arab schools gives them inadequate preparation for the critical and creative thinking required at college or university.[47] In spite of the barriers, Israeli Arabs have taken advantage of the big expansion of higher education in Israel that got underway in the 1990s, with their share of the Israeli student population pursuing a bachelor's degree increasing from 7% in the 1995–1996 school year to 13% in 2014–2015. This figure is far below their 20% share of the population,[48] but it also understates the actual share of Israeli Arabs in higher education because many opt to study in the West Bank or Jordan where coursework is in Arabic and the campus culture more familiar. By one estimate, some 8000 Israeli Arabs study in Jordan every year and another 1300 in the West Bank,[49] which numerically is a significant addition to the 18,000–20,000 studying for their undergraduate degree in Israel and closes much of the gap with the Jewish population. But it also exacerbates Israeli-Arab apartness from mainstream Israeli society, reinforcing their linguistic difference, and depriving them of the social networks and student-job opportunities they stand a better chance of obtaining attending an Israeli university.

In any case, the growth of the Israeli-Arab student population is highly skewed across several parameters, which to a degree undermines the achievement. Graduation rates for Israeli Arabs in undergraduate programs have been declining, with Israeli Arabs accounting for

9.4% of the total in 2009–2010, down from 10.7% five years earlier.[50] The percentage of Israeli Arabs who drop out or take longer than the ordinary three years are much higher for Israelis Arabs than for Jews—15.4% versus 10.8% in the 2008–2009 year.[51] Under the circumstances, it should be no surprise that far smaller proportion of Israeli Arabs pursue advanced degrees than their Jewish peers: At the master's level they accounted for only 10% of all students and at the doctorate level just 5.2% in the 2014–2015 year.[52] At the undergraduate level, the progress Israeli Arabs have made over the last three decades has been uneven: Women have been leading the growth in higher education for Israeli Arabs while men lag. The female share of the Israeli-Arab undergraduate population grew from 40% in the 1990s to 66% today, a much higher share than the 54% among Israeli Jewish women.[53] In an economy whose job market offers the highest pay and best career prospects for science and engineering skills, a far lower percentage of Israeli Arabs pursue degrees in relevant fields; in the 2014–2015 year, only 13% were studying engineering or architecture, compared with 21% of Jewish students.[54] Equal percentages were studying math and science (11%), but large numbers of Israeli-Arab graduates fail to find jobs in the fields. An interesting exception is that Israeli Arabs are overrepresented relative to Jewish students in medicine and allied health professions: Some 2% of them are studying to become doctors (double the Jewish rate) and 15% are in related programs, triple the Jewish rate.

This breakdown to a large extent reflects the realities of the job market for educated Israeli Arabs. As a rule, Israeli Arabs are less likely to be working in the field in which they were trained, but the odds are particularly poor in engineering and science, where employment is dominated by big companies and start-ups.[55] Ironically, however, the inability of Israeli-Arab graduates in math and science to find jobs in the private sector has forced many of them to turn to teaching, enabling Arab schools to increase the number of hours devoted to math and science at a time when Israeli Jewish schools are struggling to find teaching staff in these fields.[56] In turn, this may explain the growth in Israeli-Arab students pursuing degrees in science at the tertiary level. By contrast, with engineering and science, the odds of finding employment in law, accounting, and medicine are better because a graduate can go into a private practice in the Israeli-Arab community or work in the largely state-run health system, which unlike most of the civil service welcomes Israeli Arabs.[57] The above figures don't include students in teachers colleges, where Israeli-Arab undergraduates—in particular women—are disproportionately concentrated.[58] Here, perhaps is the biggest problem facing educated Israeli Arabs, because they are tracked into a field where the employment prospects are effectively limited to schools in the Arab sector.

Why Israeli Arabs are in an inferior position to Israeli Jews by virtually every socioeconomic measure is the subject of a debate that views the problem from two poles—one of that holds Israeli society and the government principally responsible and the other that assigns blame mainly to Israeli Arabs themselves. Certainly elements of both come into play; the difficulty often is deciding which of them is the fundamental cause. For instance, Israel's government consistently allocates less money to Arab local authorities than it does to Jewish ones and, despite a series of programs over the last two decades to at least partially rectify the problem, the funds that are made available end up never being fully allocated.[59] On the one hand, the phenomenon could be ascribed to a dismissive attitude on the part of officials toward the Arab minority borne of racism and to a bureaucracy that counts very few Israeli Arabs in its ranks. On the other hand, the blame could be placed on Israeli-Arab political leaders, who on a municipal level are often poor managers and on a national level don't engage aggressively in practical, interest-based politics that other ethnic parties in Israel do.[60]

Even if the line between racism and other factors is difficult to draw, the evidence of racist attitudes by Israeli Jews and instances of institutionalized racism is abundant. The racist attitudes are captured in the surveys noted above that show a reluctance to engage in personal interactions with Arabs, whether they are fundamental ones like intermarriage or the ordinary act of seeing a doctor. Some of these attitudes can be explained by the fact that religious practice, or lack thereof, is so important a badge of identity that even among the various segments of Jewish society there is little intermarriage or friendships across groups.[61] However, Israeli Arabs face discrimination of the kind that the various sub-groups of Jewish Israelis don't. Asked in a Pew Research Center survey about their personal experience with discrimination, between 15% and 17% of Israeli-Arab Muslims reported they had been questioned by security officials, prevented from traveling, or physically threatened or attacked because of their religion in the previous 12 months, while 13% said they had suffered property damage.[62] All in all, 37% of Muslims said they had experienced at least one form of discrimination due to their religion in the previous year. The security-related instances of discrimination are rooted in Israel's conflict with the Palestinians, but as noted before this easily spreads to attitudes that are expressed in vandalism or personal violence directed against Israeli Arabs. It also extends on an institutional level to legislation, where a series of laws approved in recent years by the Knesset that with greater or lesser justification have been interpreted by Israeli Arabs as discriminatory. These include the 2010 citizenship law that only requires non-Jews seeking citizenship to make a pledge of allegiance, or rules giving Israeli army veterans preference of civil service jobs, legislation allowing "small communities" the right to reject applicants incompatible with their "social fabric," and the Nakba law that prohibits public organizations from receiving government funds if they depict Israel's founding as a day of mourning.[63]

The job market, in particular, gives evidence that Israeli Arabs face institutionalized discrimination. Although there are a host of objective reasons explaining at least in part the failure of Israeli Arabs to fully integrate, survey data collected by government make clear that they also contend with discrimination based on negative attitudes. Among various groups prone to discrimination in Israeli society, a 2013 Industry Ministry survey found that Israeli Arabs were the second most likely to report being rejected for a job based on their status and were the most likely to report they faced discrimination at their place of work.[64] Among all workers, 48.3% said they perceived Israeli Arabs to suffer severe or very severe discrimination, the third highest rate after mothers of young children and the handicapped.[65] These results should be read with caution because the majority of Jewish Israelis don't share a workplace with Israeli Arabs at all, so that their perceptions aren't based on their personal experience. Even in the case of Israeli Arabs' reporting personal experiences of discrimination they are reporting perceptions, not proven facts. However, an inescapable figure is that 29.9% of the survey's respondents said they would not like to see more Israeli Arabs in their place of work, by far the highest rate among the discriminated groups in the Industry Ministry survey.[66] In this context, it follows that even when they have the required training and education, Israeli Arabs are less likely to be in jobs appropriate to their skills and when they do find appropriate employment they are more likely to work as self-employed professionals on the assumption that neither the government nor a large company would employ them.[67]

Discrimination—or at least feelings of discrimination—is a deterrent to Israeli-Arab entrepreneurialism. One entrepreneur who mills coffee launched an ad campaign for his products touting them as authentic "Arab coffee" only to discover that both Jewish and Arab shoppers shunned it on the assumption that any product manufactured by an Arab business would be inferior to one made by Jewish Israelis.[68] In fact, in the food industry where ethnic products should appeal to Jewish consumers who have adopted indigenous foods like hummus with enthusiasm, few Israeli-Arab companies advertise in Hebrew to attract Jewish-Israeli customers.

It would facile, however, to attribute the socioeconomic gaps between Israeli Jews and Arab exclusively to discrimination. The fact that Israeli Arabs are in so many respects apart from the Jewish majority by virtue of language, religion, and geographical separation influences the socioeconomic profile of Israeli Arabs. How much of a weighting to give them is a complicated matter, but one area where these differences manifest themselves is in the especially low rate of employment among Israeli-Arab women. This has been attributed to the lack of day care and public transportation in Arab towns, which can be traced back to government neglect and more deeply to discriminatory attitudes by the Jewish majority. But it is also influenced by attitudes toward the role of women in the family and society that differ from Israeli Jews. In a 2005 survey, 24% of Israeli Arabs said it was sufficient for one spouse (presumably the husband) to be working, double the rate among Jewish respondents; 97% of Israeli Arabs said women should work less to devote more time to their young children, compared with 71% of Jews.[69] Israeli-Arab women join the labor force at about the same low rate as their sisters in other Arab Muslim countries.[70] Religion, not simply an amorphous featly to traditional family structure, seems to be a factor in these attitudes: Among Christian Arabs, for instance, 34.4% of working age women are in the job market, which is far lower than the Jewish rate but more than double the 16.8% rate for Arab Muslims in 2004.[71] The low labor force participation rate for Arab women has broad socioeconomic implications: Because it means far fewer families have two breadwinners, which lowers incomes and increases the incidence of poverty, thereby narrowing the prospects for the next generation to advance in terms of education, work, and incomes. Thus, phenomena that might otherwise be ascribed to exclusively to discrimination also have roots inside the community.

These cultural differences have narrowed since 1948 by virtue of the transition of Israeli Arabs from a tradition-bound, agricultural society to one more closely approximating mainstream Israeli society. Rising levels of education, falling fertility rates, and increased numbers of working women are evidence of that. However, this transition has been held back ironically because Israel has given the Arab minority a high degree of cultural autonomy rather imposing assimilationist policies. Arabic is an official language of the state, public schools in Arab communities teach in Arabic, and an Arabic-language media (including state-owned media) exists side by side with Hebrew media. All of this is facilitated by the fact, noted above, that Israeli Arabs mainly live in their own towns and even when they live in mixed cities they have their own neighborhoods. For personal status issues, they are subject to Muslim and Christian religious law. Israeli Arabs aren't required to serve in the army—that great assimilation machine of Israeli society—and, except for the Druze and Circassian minorities, choose not to. On the one hand, autonomy has probably served to mitigate social tensions and, with a few isolated exceptions, violent opposition to the state; on the other, it has created a parallel society, cut off from mainstream Israel and all the advantages it can offer. Many Israeli-Arab political leaders and social activists would like to expand this autonomy further, but from a socioeconomic point of view, that would almost certainly widen the gaps that already exist.

Even if Israeli Arabs are an economic underclass by every measure, they don't suffer many of the social pathologies of broken families and high levels of crime that often characterize impoverished communities. Israeli Arabs are more likely to be live in traditional nuclear families of two parents and children (64% of the population) than Israeli Jews (45%); 96% of Israeli-Arab children under age 15 live in a nuclear family, as against 89% of Israeli Jewish children. The rate of single-parent households is about the same between Israeli Arab and Jewish populations,[72] but divorce rates among Israeli Arabs are much lower at about 7.4 per 1000 married couples, versus a Jewish rate of 9.6–9.7.[73] And, while poverty rates among Israeli Arabs are much higher than for Israel's overall population, the percentage of Israeli-Arab poor in chronic poverty is no different than for the overall population, or 66%.[74] Crime statistics draw a more complicated picture. The inescapable fact is that rates among Israeli Arabs are in most categories double that of Israeli Jews,[75] although the rate of family violence is much lower.[76] However, the crime rate in Israel in general is low, and the high rate among Israeli Arabs should be measured against the fact that as a group they are poorer and younger than Israeli Jews, two factors that all other things being equal would raise the crime rate.[77] As well, policing in Arab towns is poorer than in Jewish ones. All this suggests that Israeli Arabs—much like ultra-Orthodox Jews, as discussed in Chapter 11—have a firm social foundation for being easily integrating into the workforce and taking advantage of educational opportunities. On a very preliminary basis, this has been in evidence in the sharp drop in the Israeli-Arab poverty rates in 2012–2013, which occurred because of a rapid movement into the job market.[78]

While adherence to traditional social mores plays an important role in enabling Israeli Arabs to avoid many of the social consequences of poverty, they also create barriers to integrating into the labor market and in particular into Israel's knowledge economy. One small example of how strong family units may serve as a deterrent in the higher level of intergenerational support among Israeli Arabs versus Jews. A 2005 social survey found that only 15.3% of Jews reported giving financial support to their parents, compared with 25.2% of Israeli Arabs.[79] This family solidarity on the surface seems admirable, but it may go a way to explaining why Israeli-Arab men are able to opt to leave the job market much earlier than Jews. While Arabs typically work at physically demanding jobs, such as in construction, the fact that they can rely on family financial support (not to mention government allowances) presumably lowers the threshold at which men are prepared to stop working, a phenomenon that like women's low labor force participation rate, has the effect of lowering family incomes. Adherence to traditional values has also been a factor in deterring Israeli Arabs from entering the high-tech sector.

For better or for worse, many of these traditional values have come under pressure. Even as political and intellectual leaders in the Israeli-Arab community advocate greater autonomy and Israeli Arabs as a group increasingly identify themselves as something other than Israeli, a process of assimilation is underway. The push of rising educational levels and the pull of an economy that needs to integrate Israeli Arabs into the workforce will encourage the process even if it encounters headwinds created by discrimination and the barriers between Jews and Arabs created by the Israeli-Palestinian conflict. The question is whether Israeli Arabs can retain the social assets of traditional society and family structure while letting loose from those that hold back the community. A preliminary answer would seem to be they can to a large extent because Israeli society is by Western standards still beholden to traditional values, with high rates of marriage and relatively low rates of divorce. In other words, for Israeli Arabs to approach Israeli Jewish social parameters, as they have been doing, would not necessarily fundamentally undermine existing norms. The more difficult obstacle they face is Israeli-Jewish attitudes towards Arabs, which will be difficult to change so long as the country remains in conflict with the Palestinians and the Arab world.

Notes

1. Ari Paltiel et al., *Long-Range Population Projections for Israel: 2009–2059* (Hebrew) (Jerusalem, March 21, 2012), 46.
2. Hagai Levine (ed.), *The Haredi Sector in Israel: Empowerment Through Employment Integration* (Hebrew) (Jerusalem, March 2009), 9–11. Defining "Haredi" is difficult because there is no generally accepted definition. Pollsters use various criteria, such as the last place of education and area of residence. High rates of refusal by Haredim to answer surveys further complicate the issue.
3. Ibid., 53.
4. Ibid., 63.
5. Among the recent books chronicling this period: Benny Morris, *The Birth of the Palestinian Refugee Problem Revisited* (Cambridge, 2004); *1948: A History of the First Arab–Israeli War* (New Haven, CT, 2008); Yoav Gelbe, *Palestine 1948: War, Escape and the Emergence of the Palestinian Refugee Problem* (Eastbourne, East Sussex, 2006); and Ilan Pappé, *The Making of the Arab Israeli Conflict 1947–1951* (London, 1992).
6. Eran Yashiv and Nitsa Kasir (Kaliner), *The Labor Market of Israeli Arabs* (Tel Aviv, January 2014), 15–16. It should be noted that the 1948 war left parts of Mandate Palestine (the West Bank and Gaza) outside of Israel. The Palestinian population in those areas remained in their homes and were joined by refugees from what was now Israel, thus the size of the Palestinian population in the area that had been Mandate Palestine was considerably more than 18%.
7. Nohad Ali, *Representation of Arab Citizens in the Institutions of Higher Education in Israel* (Haifa-Jerusalem, November 2013), 25.
8. Arik Rudnitzky, *Arab Citizens of Israel Early in the Twenty-First Century* (Tel Aviv, November 2015), 33–44; and Sammy Smooha, *Still Playing by the Rules: Index of Arab-Jewish Relations in Israel 2013 Findings and Conclusions* (Jerusalem and Haifa, 2014), 13.
9. Rudnitzky, *Arab Citizens of Israel*; and Smooha, *Still Playing by the Rules: Index of Arab-Jewish Relations in Israel 2012 Findings and Conclusions* (Jerusalem and Haifa, 2013), 98.
10. Rudnitzky, 31–32.
11. Tamar Hermann, *The Israel Democracy Index 2015* (Jerusalem, 2015), 77.
12. Smooha, 18.
13. Hermann, 88.
14. Nadia Hilou and Idan Haim, "Civilian Service in Israel's Arab Society," *Strategic Assessment*, Vol. 17, No. 1 (April 2014), 59.
15. Ibid., 61.

16. Hermann, 74.
17. Yashiv and Kasir, 35. East Jerusalem, which has the biggest single concentration of Arabs inside Israel's border, have an unusual status: After the 1967 war, when the eastern half of the city was annexed into Israel, they were offered citizenship, but the great majority have rejected the offer and have the status of permanent residents with limited rights and benefits.
18. Ibid., 36–37.
19. Yuval Dror, *Israelis in the Digital Age* (Rishon LeZion Israel, May 2012), 10. The rate of Arabs who are connected to the Internet but don't use it is 22.2% versus 9.2% for non-immigrant Jewish Israeli. Israeli Arabs less likely to use the Internet frequently.
20. National Insurance Institute, *Dimensions of Poverty and Social Gaps, 2013* (Hebrew) (Jerusalem, November 2014), 60. Unless otherwise stated, the poverty rates refer to individuals, not to families.
21. National Insurance Institute, *2008 Annual Report* (Hebrew) (Jerusalem, June 2009), 346; and *2014 Annual Report* (Hebrew) (Jerusalem, September 2015), 354.
22. NII, *Dimensions of Poverty, 2013*, 27.
23. Yashiv and Kasir, 36–37.
24. Ibid., 19.
25. Finance Ministry, *Israeli Labor Market-First Quarter 2015* (Hebrew) (Jerusalem, undated), 9–10. Using a different age range (25–64), the participation rate for males had risen to an average of 75.1% in the four quarters through first quarter 2015. Still, that was nearly 10 percentage points less than the rate for Jewish men.
26. Myers-JDC-Brookdale Institute, *The Arab Population in Israel: Facts and Figures 2012* (Jerusalem, March 2012), 3.
27. Yashiv and Kasir, 30 and 31.
28. Yashiv and Kasir, "Patterns of Labor Force Participation Among Israeli Arabs," *Israel Economic Review*, Vol. 9, No. 1 (2011), 61.
29. Yashiv and Kasir, *Labor Market*, 34.
30. Ibid., 34.
31. Myers-JDC-Brookdale, 3.
32. Yashiv and Kasir, *Labor Market*, 34.
33. Ibid., 34.
34. Ibid., 20–21.
35. Khaled Abu Asba and Muhammad Abu Nasra, *The Palestinian Population in Israel at the Margins of the Israeli Economy* (Jerusalem, November 8, 2012), 9. The figure is from 2002.
36. Ibid., 11.
37. Ibid., 17–19.

38. Uri Shwed et al., *Integration of Arab Israelis and Jews in Schools in Israel* (Jerusalem, December 2014), 328. The authors estimated that out of a nationwide school population of 1.6 million in 2013, only 5000 Arabs and 177 Jews attended schools where 1% or more of the population was Arab in Jewish schools or vice versa. It should be stressed that there is no legally enforced segregation.
39. Dattal Lior, "TheMarker Investigation: In Secular Schools, Classroom Crowding Is Twice That in Religious Schools" (Hebrew), *TheMarker*, August 11, 2015.
40. Michal Belikoff, *Gaps Between Jews and Arabs in the Educational System—Physical Infrastructure* (Hebrew) (Jerusalem, November 2014), 11, 16.
41. Meirav Arlosoroff, "Ignoring Arab Education Imperils Israel's Future," *Ha'aretz*, March 11, 2014.
42. Israel Education Ministry, *Meitzav 5775 Part A, Achievement Tests* (Hebrew) (Jerusalem, November 2015), 32. The tests are for English, math, science/technology, and language.
43. Israel Education Ministry, *Initial Results of International Research—PISA 2012* (Hebrew) (Jerusalem, December 3, 2012), 6.
44. Ibid., 14.
45. Azulay et al., *Facts and Numbers 2016* (Hebrew) (Jerusalem, 2016), 61.
46. Central Bureau of Statistics, *Statistical Abstract of Israel, 2012* (Jerusalem, 2013), Table 8.20, *2013*, Table 8.26, and *2016*, Table 8.26.
47. Ali, 25.
48. Israel Council for Higher Education, *The Higher Education System in Israel 2015* (Hebrew) (Jerusalem, March 2015), 39.
49. Ali, 24.
50. Ibid., 21.
51. Ibid., 26.
52. *The Higher Education System in Israel 2015*, 40.
53. Ibid., 39–40.
54. Ibid., 41.
55. Sami Miaari et al., *Occupational Mismatch Among College-Educated Arabs in Israel* (Hebrew) (Jerusalem, November 5, 2012), 1, 11.
56. Central Bureau of Statistics, "Education Indicators for Training Professionals in Science and Technology in Israel" (Hebrew) (January 14, 2014), 2. The number of teaching hours devoted to science, technology, and math in Arab schools rose 74% between the 1999/2000 and 2009/2010 school years (latest figures available) while it declined almost 5% in Jewish schools.
57. Ibid., 2. In 2009, only 7% of the Israeli civil service was Arab.
58. Ali, 20.
59. Rudnitzky, 74.

60. Ibid., 75–76. The activities of Israel's ultra-Orthodox parties make a good point of comparison. Ideologically, they are hostile to the idea of a Jewish state but largely ignore that in order to fully participate in the political process, including joining governments, in order to ensure their financial and religious interests are served.
61. Neha Sahgal and Alan Cooperman (eds.), *Israel's Religiously Divided Society* (Washington, March 8, 2016), 28. For instance, 90% of secular Israelis say all or most of their friends are also secular Israelis, although only half of Israelis define themselves as secular. Among other Jewish groups, the rate varies from 48 to 89%.
62. Ibid., 31.
63. Rudnitzky, 76.
64. Shuki Handels, *Feelings of Discrimination*Discrimination *Among Workers and Job Seekers and Diversity in Workplaces* (Hebrew) (Jerusalem, February 2013), 24, 26. The survey was taken in December 2012/ January 2013 among 1050 people. Among Israeli-Arab respondents, 28.7% said they were rejected for a job because of the status, exceeded only by the 36.2% for handicapped. Among those employed, 21.1% of Israeli Arabs reported feeling discriminated against, exceeding the rates for mothers of young children and the handicapped.
65. Ibid., 27. Moreover, the percentage has grown from 39% in 2009 to 42.3% in 2011.
66. Ibid., 20–21.
67. Miaari et al., 1.
68. Abu Asba, 17. Information based on a survey of eight Israeli-Arab food makers.
69. Yashiv and Kasir, "Patterns," 58.
70. Ibid., 66.
71. Eran Yashiv, *Arabs in the Labor Market in Israel* (*Economic Debates, Second Series*) (Hebrew) (Jerusalem, January 2012), 13.
72. Bureau of Statistics, "Family Day—Families and Households in Israel," February 7, 2016, 1–2.
73. Central Bureau of Statistics, Divorces—Selected Data.
74. NII, *Dimensions of Poverty, 2013*, 39.
75. Knesset Research and Information Center, *Data on Crime in Arab Society in Israel* (Hebrew) (Jerusalem, February 23, 2010), 3. For example, in 2009, Israeli Arabs accounted for 41% of all suspects in homicide cases, 36% of those attempted homicide cases, 36% in cases of aggravated assault, and 36% in cases of robbery, even though Israel Arabs comprise 20% of the total population.
76. Ibid., 4. In all likelihood, the rate of reporting of family violence in the Israeli-Arab community is lower than for the Jewish community but

given the rate of the gap, violence the rate is probably not that high even including unreported.

77. Ibid., 6.
78. Ibid., 27.
79. Yashiv and Kasir, "Patterns," 69.

CHAPTER 11

Outsiders II—The Ultra-Orthodox

As victims of discrimination rooted in their otherness, Israel's Arab minority is not untypical of social underclasses around the world. Although they are to a degree constrained by their own social values, culture and a desire to preserve their separate identity, Israeli Arabs' inferior status in the main is imposed on them. Given the opportunity, they have gladly exploited educational and job opportunities to improve their position. By contrast, most of Israel's ultra-Orthodox (Haredi) community ethnically speaking belongs to Israel's traditional elite of Ashkenazi Jews. But Haredim have chosen as a community to isolate themselves from Israeli society, shunning the world of work and mainstream values in favor of a life of strict religious observance and study and—as an inevitable consequence—taking on the burden of severe poverty.

The result is a confusing and often contradictory maze of socioeconomic parameters. Measured by years in school and resources devoted to education, Haredim are highly educated. But their studies are focused almost entirely on religious texts that provide no foundation for life in the modern world, much less work in a knowledge economy. In a reversal of typical sex roles, Haredi women are more likely to receive a general education than Haredi men. Yet this advantage earns them no economic power or leadership role in the community, nor do they openly aspire to them. Apart from child-bearing and rearing, a woman's primary role is to earn money to support the family, and as a corollary, she needs schooling to prepare herself for the job market. But in the Haredi world, work is regarded as inferior to religious study and is left to women, whose status

D. Rosenberg, *Israel's Technology Economy*, Middle East in Focus,
https://doi.org/10.1007/978-3-319-76654-6_11

is inferior to men. Despite high rates of poverty in the Haredi world and a heavy reliance on government allowances and private charity, the social controls imposed by religious law, tradition and rabbinic leadership mean that Haredi society exhibits few of the social maladies that typify impoverished communities. Family and community life are strong, and crime rates are low.

The Haredi relationship to the modern world is laden with contradictions. On the one hand, the men dress in black and often in style that harks back to nineteenth-century Eastern Europe (coincidentally, giving them a similar appearance to American Amish). Haredim shun the Internet and other forms of modern media and communications. On the other hand, they have no objection to technology per se and compared to fundamentalist Christians do not see rejection of scientific theories like evolution and climate change as a critical test of faith. The ultra-Orthodox are concerned first and foremost with the minutiae of Jewish religious law, not with the theology that underpins it. The community distinguishes itself from other Israelis and from other Jews principally by practice and only secondarily by ideology. Thus, the characteristics of Haredi Judaism identified by Menachem Friedman are framed primarily in terms of the daily life of the community and not by beliefs.[1]

The first group of these characteristics relates to the essence of Haredi community life, namely its complete religious faith, which in practical terms is defined as strict adherence to Jewish law (*halacha*) and tradition. The daily routines of ultra-Orthodox Jews are dictated by an elaborate and all-encompassing system that defines, among other things, what is permissible to eat, strictly regulates times for prayers and other observances, punctuates the year by Sabbaths and holidays where ordinary, everyday activities like using electricity are banned, establishes severe strictures on what constitutes "modest" dress and regulates sexual relations between husband and wife as well as day-to-day contact between men and women generally. The system is predicated on a belief system that gives divine imprimatur not only to halacha as it appears in the Bible and Talmud, but also to a vast corpus of religious literature written over the centuries, traditions and the contemporary rulings (*psakei din*) of rabbis—a body of writing that comes under the rubric "Torah." It assumes God's active presence in the world, although it stops short of imposing a doctrine of fatalism over day-to-day life.[2] Where ordinary Haredim draw this line between free will and divine intervention is an important factor in terms of their ability to integrate into a knowledge society that is based on the idea of personal autonomy and the ability of people to effect change and control their lives. While one survey found nearly 78% of Haredim said they believed divine providence was the main factor determining a person's social status,[3] Haredim, for instance, readily accept medical treatment and have shown a capacity for business enterprise that suggests they tacitly accept the capacity of science to overcome fate and of personal initiative to bring economic benefits. Thus, on the balance, the Haredi worldview doesn't seem to be an obstacle, which may reflect the paramountcy of religious law and practice over theology.

The second group of characteristics relates to the community's all-encompassing social structure and strict fidelity to its rabbinic leadership. In contrast to other streams of Judaism, the role of rabbis of the ultra-Orthodox world is paramount: They not only provide spiritual guidance and serve as educators, but the leading rabbis of the community are regarded as "Torah luminaries." These figures' exalted status bestows on them the authority to make decisions for their followers on matters like marriage, medical treatment and business that don't involve halacha per se and gives them the authority the community's political leaders.[4] Their status is not derived from their learning, which is a given, but from the charismatic power and deep understanding of earthly matters ascribed to them, which gives their decisions the force of divine inspiration (*da'at Torah*).[5]

The third group of characteristics is connected with the centrality of education—the principle that Jewish males should ideally be engaged in religious study most, if not all, of their lives; conversely, ordinary work of any kind is looked as at best an inferior to study and at worst a potential handicap to a proper Jewish life. The Haredi world is organized as a "Society of Learners," where not only male children and adolescents are engaged in full-time study but so have been in recent decades the majority of adult men (*avrechim*).[6] The yeshiva and kollel, the core institutions of religious learning, stand at the center of ultra-Orthodox life, serving not just as an educational institution but as a bulwark of ultra-Orthodox values and practices.[7]

The fourth group of characteristics defines the relationship of the ultra-Orthodox community to the world, which is characterized by opposition to all outside culture, lifestyles and values. In Israel, this includes anti-Zionism, a refusal to serve in the army and a unique style of dress that sets them apart from others.[8] In the most extreme circles,

Haredim prefer Yiddish instead of Hebrew as their day-to-day language and refuse to recognize or cooperate with state institutions, including voting or paying taxes, or conversely, receiving state benefits.

These values are subject to change over time and are not shared equally across the Haredi world. The ultra-Orthodox community in Israel contains three major streams, whose attitudes differ to one degree or another. For instance, the Lithuanian stream, which comprises an estimated 29% of the Haredi population,[9] takes an extreme attitude toward the role of learning and the spurning of men working for a livelihood.[10] Although its ideology has in recent decades spread to other parts of the ultra-Orthodox world, differences remain. Another third belong to the Hassidic stream, which began as a spiritual-revival movement in eighteenth-century Europe and to this day takes a more mystical approach to Jewish practice. On the whole, the Hassidic stream does not encourage its followers to engage in a lifetime of full-time study, except for an elite minority.[11] Some of the 100 Hassidic courts (sects) even permit their members to serve in the army and look on Zionism favorably. Finally, the Sephardi stream, which comprises much of the reminder, follows the Hassidic practice regarding work and the army.[12]

Elements of Haredi Judaism can be found throughout Jewish history, and proof texts can be offered to assert its claims to represent an authentic, timeless form of the faith. But ultra-Orthodoxy is a relatively new phenomenon in Jewish life, a counter-revolution to the forces of emancipation, assimilation and modernity that began to infiltrate the traditional Jewish world in Europe at the end of the eighteenth century and led to the creation of Reform and Conservative Judaism. In the first decades of the following century, the rudiments of what would become ultra-Orthodoxy began to take shape under the leadership of Rabbi Moses Sofer (known as the Chatam Sofer), who fought against any change in traditional Jewish practice and asserted that Torah made no allowance for change or innovation. In 1865, nearly three decades after his death, ultra-Orthodoxy began coalescing in a nine-point p'sak din that formulated the first conscious statement of ultra-Orthodoxy as an ideology (with a critical 10th point banning secular education added later).[13] Over the following decades in eastern and central Europe, the outlines of ultra-Orthodoxy were drawn, including opposition to innovation and reform, resistance to cultural assimilation, opposition to the emerging Zionist movement and the principle of Torah study as paramount.[14]

The world of ultra-Orthodoxy was profoundly shaken by the Holocaust and the creation of the State of Israel. The Holocaust quite literally put an end to Jewish life in the towns and cities of eastern and central Europe that had been its intellectual center and the home of the great majority of its adherents. Entire communities were destroyed, scholars were slaughtered, and the great yeshivot were shuttered. The center of Jewish life by default moved to the USA and Israel, where existing communities were reinforced by the remnants of European ultra-Orthodoxy that settled in them. For most of Jewry, the tragedy of the Holocaust was compensated in part by the rise of the Jewish state; for the Haredi world, they were twin tragedies. The Zionist movement and the settling of the Land of Israel were seen as a rebellion against God's will. However, the creation of the state in 1948 left the Haredim with little choice but to establish a modicum of cooperation with the new political masters of the Holy Land.[15] Although the impending state's leaders were officially committed to secular values, they aspired to national unity and were prepared to reach understandings to ensure the ultra-Orthodox world's consent. In addition, both sides acted in the shadow of the terrible destruction wreaked by the Holocaust on the Jewish world—although the conclusions they drew from the tragedy were very different. The Haredi world's rabbis determined that the task of their generation would be to rebuild European Jewish life as they remembered and idealized it; Israel's leaders acted in a spirit of tolerance and generosity toward them on the assumption that the ultra-Orthodoxy was destined to disappear in the modern era. The result was a series of agreements and laws that spelled out a modus vivendi between two sides that would effectively form the basis for the Haredi Society of Learners in the State of Israel.

The first was an understanding, which predated Israel's creation, spelling out the relations of the incipient state to Judaism. Known as the "status quo" agreement, the 1947 accord stated that Israel would not be a theocratic state but that it would designate the Jewish Sabbath as a day of rest, observe the laws of kashrut in official institutions and subject personal status issues for Jews to rabbinical courts dominated by Haredi rabbinic judges.[16] Most importantly, it granted institutional autonomy to ultra-Orthodox schools, albeit with the proviso they would teach a core curriculum of secular studies. The independence of Haredi schools was institutionalized in 1953 as part of a broader restructuring of the educational system that created three classes of schools—a wholly secular public system, a system of public religious schools for non-Haredi Orthodox Jews, and a so-called independent network of Haredi schools subsidized by the state but, as it turned out, not supervised by it.[17] The principle that Haredi men would not be drafted also predated the state, under an agreement reached in March 1948 that temporarily exempted young men who were engaged full-time in religious study ("Torah study as their occupation," as it was defined).[18] When Menachem Begin and his right-wing Likud Party won the 1977 elections and ended 29 years of rule by the Labor establishment, the number of exemptions expanded greatly. In 1948, it was a mere 400. But it grew nearly 12-fold over the next two decades to 4700 and ballooned to 18,000 in 1988, 24,000 in 1995, 45,000 in 2006, and 60,000 in 2011,[19] a rate far in excess of the growth in the ultra-Orthodox population.

The model of the Society of Learners failed in Eastern Europe because there was no means of financially supporting large number of families headed by perpetual students,[20] but that last building block of contemporary Haredi society came into place by virtue of the Israeli government's readiness to fund avrechim and their families. State funding in the form of income support and child allowances, as well as discounted government services, became the economic foundation of Haredi life starting with the Begin government. Agudat Yisrael, which ended a 25-year boycott of successive coalitions to join the Begin government, institutionalized a system of providing critically needed votes to ensure economic benefits to their constituents. Other ultra-Orthodox parties adopted the practice.

The Haredi world has not created the economic autarky inside Israel that it theoretically aspires to; to the contrary, as it has come to rely on the state for allowances and even employment, it has become a dependent. But its economic life, like its social and political life, is very different from the society that surrounds it. At its center is the educational establishment, whose importance is not only due to the community's comparatively large population of children but a large proportion of the adult male population. In 2011, the Haredi schools comprised 28.9% of Israel's entire elementary school-age population, a rate more than three times its share of the overall population.[21] Yeshiva schooling occupies a far larger part of a child and adolescent's waking hours than for his or her secular peers. The study of Torah is regarded as the ideal of Jewish life.[22] One reason for this is that ultra-Orthodox educational institutions are not as simply transmitters of knowledge but a conduit for ensuring that the practices, values, and worldview of the community are handed down to the next generation in a highly controlled environment.[23] Given the imbalance of forces between the Haredi community and mainstream Israel in terms of sheer numbers, economic and political power, and the influence of the media, Haredi educational institutions see an important part of their task as keeping every aspect of the outside world at bay. In fact, as it is structured, the yeshiva—the central institution of Haredi education—has taken over the traditional role of the family of ensuring continuity in religious practice and belief over the generations.

For boys, formal education usually begins at age three in nursery schools and moves on the kindergarten for ages four through six.[24] The equivalent of elementary school education for ages 6–13 is the *talmud-torah*, which provides what for most boys will be all the general education of arithmetic, language skills, and the like they will ever receive—and even then it comprises a small and neglected part of their studies. The next critical step is when boys move to a *yeshiva k'tana*, where their days are devoted solely to religious studies through age 15 or 16. From 16 on, they study at a *yeshiva g'dola* until marriage, which is typically in their late teens or early twenties. School days even at a talmud-torah level are long, but in the yeshiva g'dola education becomes all-encompassing: Boys typically enroll at an institution where they both live and study, putting them under the supervision of the yeshiva rather than their families. Whereas in other Israeli elementary schools, classes in math, English, and science are required, in Haredi institutions the rates at which they are taught range between 83% for math and 53.6% for English.[25] At the post-elementary level, the rates fall further to 41.3% for math and 38.5% for English. Given the disproportionate size of the Haredi school-age population in Israel, this has a profound effect on the overall number of Israelis getting a general education. For instance, only 85.3% of the entire Israeli high school-age population has any access at all to math courses and only 62.4% to computer science (in fact, more Israeli Arabs have access to Hebrew language classes in high school than do Haredi Jews). This is a severe handicap to an economy whose future hinges on high technology.

The only reason the rates for general education in Haredi schools are as high as they are is due to the fact that the great majority of girls study in the Beit Yaacov network of schools, where they are taught secular subjects side by side with religious studies.[26] A girl's education is designed to teach her the essentials of Jewish practice but also to impart the practical learning and formal qualifications she will need to enter the labor market and support her future husband, who ideally will be an avrech dependent on her income.[27] Girls' education follows a trajectory more or less similar to non-Haredi Israelis, starting with nursery school and kindergarten, and then moving on to elementary school. Unlike boys, girls' general education continues after elementary school, enabling them to qualify them for the official matriculation certificate (*bagrut*) or a similar one called the Szold. For women over 18, the primary form of post-secondary education is Haredi teacher seminaries whose course of study qualifies them for a teaching certificate.[28] More recently, young women have begun completing courses in computers and other professions that offer a wider assortment of job opportunities.

Before the creation of the State of Israel, funding for Haredi educational institutions came principally from the private donations of diaspora Jews, who were not religiously observant themselves but saw such yeshivot as "museums" of the Jewish life their parents and grandparents led and contributed to their upkeep out of a sense of nostalgia or Jewish identity.[29] Private funding, however, was insufficient to build and operate an educational system on the scale that a Society of Learners would require. The 1953 law formalizing the structure of the Israeli educational system, which included the Haredi "independent" stream, changed that, enabling funding for ultra-Orthodox schools to grow massively.[30] Today, Haredi school budgets are covered by the government at a rate that varies from 55% to 100% of what other schools get, depending on the degree to which they teach a mandated core curriculum of general education.[31] But even the schools qualifying 100% funding are not inspected on a regular basis by the Education Ministry to see if they are in compliance, and evidence from standardized exams (in the few ultra-Orthodox schools that consent to letting their students take them) indicates that what is being taught is being taught poorly.[32] In any case, formal and anecdotal evidence suggests that the general education curriculum is taught in isolation, with little or no context, as an appendage to the main program of religious studies; it has little bearing on the students' real concerns or interests,[33] so that its impact of children's and adolescents' lives and ways of thinking is marginal.

The practice of denying Haredi males a general education in favor of an exclusively religious education only took hold in the 1950s at the prompting of Rabbi Avraham Yeshaya Karelitz, the community's leader in the 1930s and 1940s and popularly known as the Chazon Ish.[34] But it is regarded as an age-old Jewish ideal, and the Haredi world is deeply committed to it even as the ultra-Orthodox population has grown, making the absence of Haredi men from the labor market in significant burden on the rest on the ultra-Orthodox and general Israeli economy. The Education Ministry adopted an official core curriculum, including math, science, and English, in the 1990s, and the High Court of Justice ruled in 2000 that school funding must be contingent on Haredi schools teaching it, but the standards were never enforced due to pressure from Haredi political parties. When they were briefly out of the government, the Knesset passed a law in 2014 denying funds to schools that fail to teach the core curriculum starting in 2018, but Haredi parties succeeded in getting the law reversed two years later after they rejoined the coalition and the status quo remains as of this writing.

The life of learning for Haredi men continues into marriage and adulthood, which is the key to understanding the workings of the Haredi economy. Not all males study full-time in a kollel, the institution for married men continuing their studies, but a government study estimated that the figure in 2009 reached as high as 85% for men aged 20–24, and while it trends down as men grow older, in the prime working age of 25–34 about 75% were in kollelim.[35] Into their late thirties and mid-forties, over half were still in full-time study, and for those aged 50–55, 28% were, meaning that they will have spent nearly their entire adult life outside the labor market. The phenomenon has both an ideological basis and a practical, socioeconomic one. Ideologically, Haredim for the most part don't see kollel study as having any utility, not even a religious one in the sense of qualifying the avrech for employment in religious services. In its purest form, study is regarded as a religious act in its own right, akin to prayer or observing the Sabbath, while at the same time providing a spiritual defense of the Jewish people and the State of Israel.[36] But there is a practical purpose, from the community's point of view because kollel study exempts avrechim from army service, which the community fears would expose young men to the temptation of the non-Haredi world. Kollel study also has the effect of preventing a young man from entering the labor market even on a part-time basis because the government stipends for avrechim are contingent on full-time study. Another practical consideration is that in the ultra-Orthodox world, where marriages are arranged by families in accordance with practical and ideological imperatives, an avrech has higher social standing than a working man, regardless of the latter's earnings or future earning ability, and has better prospects of finding an ideal mate.[37]

Haredim hold that scholars will be sustained financially by miracles, but in the twenty-first century the economic basis of the system is government benefits of a size and scale that only Israel offers kollel scholars and thus hosts the only significant ultra-Orthodox community where full-time study is so pervasive. In Britain, for instance, where avrechim get no state support, the percentage of men in kollelim never exceeds 36%—and then only for the youngest adults.[38] By their mid-thirties, the rate falls into the single digits. In Israel, kollel families' incomes include child allowances, income support and Education Ministry grants.[39] This state assistance is typically supplemented by income from the avrech's wife and private subsidies from the kollel and community. Government benefits are generous enough that by one government estimate, a family sustained by a working wife, together with state and private support, can earn as much as a family with two working breadwinners.[40] After factoring the extra childcare costs of a two-income family, even an avrech and a non-working wife are not much worse financially off than a working couple. In any case, a Haredi male who might consider employment is unprepared in terms of education or experience to hold a job lucrative enough to justify leaving the kollel.

Against the seemingly insurmountable obstacles raised by their pre-tertiary education and by community disapproval, more Haredim have begun pursuing a higher education in recent years, in part due to a concerted government program called *Machar* that offers sex-segregated classes and preparatory programs. In the eight years from the 2005/2006 school year to the 2013/2014 school year, the number of ultra-Orthodox students in institutions of higher education rose from about 3000 to 8050 and was closer to 10,000 in the final year, counting Haredim who study in ordinary classes rather than in the Haredi framework.[41] In any case, the phenomenon should be kept in perspective. The number of Haredim in higher education is still tiny, and the rate of growth has been slowing in more recent years. Evidence suggests that far from representing a new and growing norm, they are outliers: For instance, only 25.3% of the female Haredi students are married, less than half the rate for Haredi women aged 20–24 as a whole.[42] At one institution that surveyed its students, 24% of its male Haredi students had served in the army, another 14% were in a special army program for students pursuing a degree before service, and 5% had done civilian national service, rates that are much higher than the rate for Haredi men as a whole.[43] Most Haredim study practical subjects like law and accounting. Far fewer than the general population pursue science, technology, engineering or math,[44] and virtually none opt for a liberal arts degree.[45]

Between segregated classes and the course of studies they choose, the higher education Israel's ultra-Orthodox are getting does little or nothing to influence their fundamental worldview or bring it appreciably closer to mainstream society. In any case, large numbers never complete their degrees. Although the ultra-Orthodox insist that a traditional religious education equips them with the basic intellectual tools that will enable them to master math and English later in life, most Haredi students struggle to cope and about two-thirds drop out in the university-preparatory stage, even though they represent a motivated and self-selected group.[46] When they reach the college classroom, 50% of men and 30% of women drop out, much higher rates than for other groups.

The refusal of Haredi men to perform military service has met with growing anger on the part of mainstream Israel over the last two decades as the number of exemptions has grown. The debate is usually framed in terms of equalizing the burden, i.e., of requiring Israel's ultra-Orthodox citizens to meet the same obligations to the state as others. However, army service has a potentially important economic dimension as a conduit for integrating Haredim into Israeli society by putting them into regular, ordinary contact with non-Haredi Israelis, teaching them job skills and the Israeli values as described in Chapter 8. Army service as a catalyst for employment and integration is evidenced by figures on employment for the tiny minority of ultra-Orthodox who have served. Seeking to address Haredi concerns about the danger of army service, the Israel Defense Forces has created Haredi-only units that cater to their special needs. In one such unit called Shahar, where soldiers provide technical support, rate of employment among veterans is 70%, which is lower than for ordinary IDF veterans but much higher than the 41.5% for all Haredi men at the time of the survey.[47]

Like the core curriculum, the draft is seen by the Haredi community as a fundamental threat to its existence, although many prefer to cast their opposition in the more positive light of Torah study providing a spiritual defense of the Jewish state that complements the army's material defense.[48] The half-century-old blanket exemption for Haredi men was disallowed by the High Court of Justice in 1998, forcing the Knesset after much political wrangling to approve what was known as the Tal Law.[49] The law formalized the exemptions process, redefining the standard from the amorphous "Torah study as their occupation" to hours devoted per week to study and other explicit criteria. The law, envisioned as a stopgap before a truly equitable system could be put into place, was supposed to lead to declining exemptions, but it had the perverse effect of making it more difficult to leave the kollel by formally barring avrechim from activities such as professional training or part-time employment. The law's failure in that regard caused the High Court in 2012 to rule the Tal Law unconstitutional, leading to legislation two years later that imposed criminal penalties on those who don't perform military service and on the institutions that enroll them. However, the Knesset rolled back the law in 2015 and the situation remains a flux as of this writing.

The emergence of an anti-labor ideology combined with government financial support for avrechim caused the labor force participation rate for ultra-Orthodox men in Israel to drop steadily over the 25 years to 2008. In the early 1980s, the employment rate for men aged 35–54 had been about 70%, which even then was 20 percentage points less than for non-Haredi Jewish males.[50] From then on, the rate dropped rapidly to below 40% by 2000 and continued to edge lower still in subsequent years. Although there has been no fundamental change in Haredi ideology, over the last decade the employment trend has reversed. Using a wider age range, the National Economic Council estimated that in 2002, only 39.2% of ultra-Orthodox men aged 25–64 held jobs or were actively seeking one, but by 2015 the figure has reached 53.7%.[51] Still, that was far lower than it had been 35 years earlier and the 90.8% rate for non-Haredi Jewish men. Among women, the labor force participation rate grew even faster—from 49.2% in 2000 to 77.6% in 2015—bringing their rate close to the 84.3% for non-Haredi Jewish women. The reasons for this rise are manifold, including reductions in government allowances after 2002, financial pressure, a growing consumer culture, and government programs formalized in 2010 to encourage Haredim to join the labor force being among them. But, as the figures show, it was Haredi women who acted on these drivers more than men. They face fewer ideological obstacles to holding jobs and are better prepared for the workforce in terms of education.

Haredi women also have the advantage of a somewhat better education—for instance, the percentage of younger women with a tertiary education was 12.8% in 2011, compared with 7.5% for men—but they are responsible for family childcare, which is a bigger burden for them compared to their non-Haredi peers, given the high average fertility rate, and means they take frequent maternity leave.[52] Employment rates among them peak in their twenties before family obligations make a holding job too difficult for many.[53] Under the circumstances, many opt for employment inside the Haredi community, where terms are more accommodating and it is easier to adhere to their religious mandates. Wages are low, but they would not earn appreciably more in the outside economy because for many employers ultra-Orthodox women are hired a low-cost, reliable labor pool that can perform relatively sophisticated tasks; for many Israeli high-tech companies, their pay is low enough to act as an alternative to outsourcing jobs to cheaper countries.[54] Close to 35% of Haredi women in the workforce were employed in ultra-Orthodox schools in 2011, more than double the rate in 1979.[55]

While Haredi women are more employment-ready than men, the employment prospects for Haredim of both sexes are very limited. Both men and women experience discrimination when they try to enter the mainstream workforce, a situation borne out by a 2012–2013 poll that found that 17.7% of respondents willingly state they opposed hiring more Haredim in their workplace.[56] Their religious imperatives, such as refraining from working closely, if at all, with members of the opposite sex, serve as an employment barrier in terms of both the range and kind of jobs available to them and companies' willingness to take hire them. But the key factor is their lack of a general education, a factor that has become a greater deterrent over time as the Israeli workforce becomes more educated and the standard for jobs rises. While the rest of Israel's population has seen its levels of formal education rise, among Haredi men it has fallen as the more and more opt for an exclusively yeshiva education. In the key ages of 35–54, the percentage of ultra-Orthodox men with only an elementary education rose to 47% in 2010 from 31% eight years earlier; those with only a secondary education shrank to 12% from 26%.[57] Among ultra-Orthodox men and women, the percentage with a tertiary degree is actually smaller for the 25-to 44-year-old cohort than it is for the 45–64 group.[58] Haredim have become a source of low-cost, low-skill labor that is undermining the movement of the Israeli economy toward high value-added work and may be a contributing factor to the economy's sluggish productivity growth.

That said, much of the Haredi employment that there is consists of providing religious services to their community as well as the great non-Haredi majority, a situation enabled by the official status of Jewish law in public institutions enshrined in the 1947 status quo agreement. Rabbinical courts, local religious councils, kashrut supervisors, institutional and army rabbis, and burial services employ Haredi Jews to serve the secular or less religious majority,[59] a market created by law, funded by taxpayers, and largely limited to ultra-Orthodox men, who are the only ones with the qualifications to fill them. Just for kashrut supervision, which is a government monopoly, the costs run to 2.8 billion shekels (about $750 million) annually, or about 3% of the food industry's total turnover.[60] Teachers make up an unusually large part of the Haredi workforce for both men and women, thanks to government budgets for their schools and the absence of alternatives. In 1979, 13.8% of Haredi men were employed in education, a figure that ballooned to 21.5% in 2011, more than five times percentage for non-Haredi men.[61] Much of this employment is artificial—not supplying a market need but based on state-enforced regulations and monopolies, and access to government funding. Its contribution to Israel's economic life is marginal and disguised unemployment. One small example of this is the increase in the number of Haredi female teachers versus the growth in the student population they serve: a 97% rise in teachers between 2000 and 2010, compared with a 65% increase in students.[62] The gap was particularly pronounced at the preschool level where qualifications are looser, making it easier for institutions to create and fill jobs.

It almost goes without saying that poverty is especially widespread in the ultra-Orthodox world, although similar to the Israeli Arab minority, it is not accompanied by high levels of social pathologies by virtue of the community's fealty to traditional values. Unlike other segments of the Israeli population, however, the Haredi poverty rate has grown in recent years even as more and more of the community enters the job market.[63] Economic pressures have forced more and more ultra-Orthodox to work, but without the skills, experience, and education for well-paying jobs, employment doesn't significantly improve their incomes. In 2013, the poverty rate was an astonishing 52.1% for Haredi families even after taking into account government allowances, nearly three times the rate for all Israeli Jews and a rate that far exceeds groups typically suffering high poverty rates such as families headed by a single parent or by someone with less than eight years of schooling.[64]

Furthermore, Haredi poverty is not in stasis in the sense that the community has aligned low income levels with commensurately low expenses, which is understandable since high birthrates raise the bar for a family to generate income that lifts them above the poverty line. A typical Haredi family manages financially by a jury-rigged system of government allowances and private charity (much of the latter from dubious sources) that falls short of its household expenses, a 2014 survey by the Taub Center for Social Studies found. Comparing five population groups in Israel by religion, it found that Haredi Jews had the second lowest monthly household income after Israeli Muslims. Income from work accounted for just 42% of the Haredi monthly income, by far the lowest among the five groups, while government allowances and private charity accounted for 26.4%, which was by far the highest.[65] The Taub study found that while virtually all population groups were running monthly household expenses in excess of their income, among Haredi families the gap was unusually wide, amounting to monthly overspending of 3209 shekels on average for the years 2007–2011 on income 9535 shekels.[66] The private charity that ultra-Orthodox families rely on for such a large part of their monthly income comes from donations by wealthy Haredim abroad and in Israel, but much of it also is financed by black capital. An important conduit for family income is aid via free-loan funds (*gemachim*). These are private charities that provide interest-free credit, of which as much as a fifth is never repaid, many of which serve as a conduit for money laundering.[67]

There is evidence that this system is unravelling. Multiple generations of impoverished families living on state assistance can't afford to maintain the traditional system of mutual assistance, most particularly parents assisting their adult children. The generation of ultra-Orthodox Holocaust survivors who received reparations from Germany starting in the 1950s is dying out, ending a key source of community income. The growth of the Society of Learners model in diaspora Haredi communities—albeit on a less intense basis than in Israel—means fewer funds are available from overseas to support the Israel community. The cost of living, in particular for housing, has risen sharply in Israel over the last decade, adding pressures on the Haredi economic system. Although small, an emergent Haredi middle class, with its aspirations for the amenities enjoyed by non-Haredi consumers, is also undermining the system by creating an alternative model to voluntary, faith-based poverty.[68]

As economic pressures have begun to chip away at the kollel ideal, access to the outside world via the Internet and more contact with outsiders has caused secular ideas and temptations to penetrate the community. Since the death of Rabbi Eliezer Schach in 2001 rabbinical authority has weakened, or at least become more diverse, so that ultra-Orthodox world is no longer beholden to a single view.[69] But the Haredi response to changing conditions is occurring too slowly and hesitantly to compensate for the rapid growth of the community's population. The Haredi community offers Israel's knowledge economy a large unexploited pool of potentially skilled labor. But the severe religious restraints that the ultra-Orthodox world imposes on itself, together with the political power of its leadership to resist institutional change imposed from above, makes the task of tapping that resource extraordinarily difficult and perhaps even impossible.

Notes

1. Menachem Friedman, *The Haredi (Ultra-Orthodox) Society—Sources, Trends and Processes*, 1–26; and Reuven Gal, *Ultra-Orthodox Jews in Israel: A Status Report, 2014* (Hebrew) (Haifa, February 2015), 2. Gal enumerates 10 characteristics from which the following section is derived: (1) Complete religious faith, (2) Extreme observance of religious law, (3) Social control, (4) Torah study as paramount value, (5) Obedience to Torah luminaries, (6) Opposition to social values outside the community, (7) Rejection of Zionism, (8) Community isolation, (9) Centrality of educational institutions, and (10) Unique dress and appearance.
2. David Landau, *Piety and Power: The World of Jewish Fundamentalism* (New York, 1993), 44.
3. Central Bureau of Statistics, Social Survey 2008, Table 33. Among secular Israelis, only 0.9% said divined providence was the main factor; among non-Haredi religious Jews, it was 38.4%.
4. Friedman, 105–106.
5. Ibid., 104.
6. Moti Bassok, "The Number of Avrechim Fell: 'The Path Is Still Long and Hard'" (Hebrew), *TheMarker*, June 21, 2016. The figures were prepared by the Israel Religious Action Center using data from the Education Ministry, which found that the percentage of adult males studying full-time was 61% in 2010 and had fallen in subsequent years but was still 50% in 2015.
7. Friedman, 7–8.
8. Ibid., 8.
9. Lee Cahaner, Nikola Yozgof-Orbach, and Arnon Sofer, *Haredim in Israel: Space, Society, Community* (Hebrew) (Haifa, November 2012), 48.
10. Daniel Schiffman, "The Ideology of Full-Time Religious Study by Married Men in Israeli Ultra-Orthodox Judaism," 4–5.
11. Ibid., 22.

12. Ibid., 23–24.
13. Michael Silber, "The Emergence of Ultra-Orthodoxy: The Invention of a Tradition," in Jack Wertheimer, ed., *The Uses of Tradition: Jewish Continuity on the Modern Era* (New York, 1991), 39.
14. Ibid., 48–83.
15. Friedman, 52–53 and Cahaner et al., 67–68.
16. Ibid., 85.
17. Ibid., 95.
18. Ibid., 85 and "Burden Sharing and the Haredim," in Shlomo Fischer, ed., *Annual Assessment 2013–2014, No. 10* (Jerusalem, 2014), 121.
19. Cahaner et al., *Haredim in Israel*, 85.
20. Friedman, 187–188.
21. Central Bureau of Statistics Yearbook, 2011, 390.
22. Friedman, 7.
23. Cahaner et al., *Haredim in Israel*, 95.
24. Reuven Gal, *How to Bring Haredim Into the Science and Technology Professions* (Hebrew) (Haifa, June 2015), 6.
25. Central Bureau of Statistics, "Survey of Educational and Welfare Services in Elementary and Post-Elementary Schools—Hebrew and Arabic Education," December 12, 2011.
26. Gal, *How to Bring Haredim Into the Science*, 6–7.
27. Friedman, 57–58.
28. Hagi Levine, ed., *Haredi Sector in Israel: Empowerment Through Employment Integration* (Hebrew) (Jerusalem, March 2009), 29.
29. Friedman, 74.
30. Friedman, 56.
31. *The Haredi Sector in Israel*, 24.
32. Lior Dattel, "Data the Education Ministry Refuses to Publish: Haredi Student Achievement the Lowest in Israel," *TheMarker*, November 8, 2012. The Education Ministry doesn't break out Haredi results in the nationwide standardized Meltzav exam, and TheMarker compared results by examining results by neighborhood. Of Haredi schools whose students participated, 44% were in the bottom decile of results and 14% in the second lowest decile.
33. Moshe Krakowski, "Worldview Construction and Identify Formation in Ultra-Orthodox Jewish Elementary Schools," *Diaspora, Indigenous and Minority Education: Studies of Migration, Integration, Equity and Cultural Survival*, December 20, 2012, examines teaching of secular subjects in American schools, but presumably the same attitudes prevail in Israeli schools as well. Sarah Liebovitz-Dar, "In the Heart a Wall: The Big Battle of the Educational System over the Future of the Country,"

Ma'ariv, May, 16, 2015, provides anecdotal evidence of how general education is taught in Jerusalem ultra-Orthodox schools.

34. Friedman, 57, 72–73 and Schiffman, 8–9.
35. *The Haredi Sector in Israel*, 49. The figures are for about 2009 and have probably fallen since then as more Haredi men enter the workforce.
36. Schiffman, 6.
37. *The Haredi Sector in Israel*, 18.
38. Ibid., 49. The figures are for 2005.
39. Ibid., 42.
40. Ibid.
41. Gal, *How to Bring Haredim Into the Science*, 12.
42. Ilia Zatcovetsky and Reuven Gal, *Between Tomorrow (Machar) and Today: The Haredi Academic System at a Crossroads* (Hebrew) (Haifa, June 2015), 21.
43. Ibid., 22.
44. Gal, *How to Bring Haredim into the Science*, 15.
45. Cahaner et al., 105–106. Approximately 90% pursue professional degrees, and of approximately 4800 students in bachelors or master's program in the 2010/2011 year, only six were studying humanities.
46. Meirav Arlosoroff, "Does Studying the Talmud Make You Smarter? Data Dispels Old Myth," *Ha'aretz*, July 27, 2016.
47. Gal, *Ultra-Orthodox Jews in Israel*, 15. The report concedes that the cause-and-effect relationship is unclear, i.e., those more inclined to join the workforce over kollel study may also prefer army service. Nevertheless, the figure does suggest that army service can overcome the educational disadvantages Haredim face in entering the job market.
48. "Burden Sharing and the Haredim," 126.
49. Ibid., 121–123 and *The Haredi Sector in Israel*, 30.
50. Ayal Kimhi, "Trends in the Labor Market: Gaps in Employment Rates and Wages," in *State of the Nation Report 2011–2012*, 3–4. Among Haredi women, where the anti-labor ideology isn't relevant, the labor force participation rate was 50% in the early 1980s and fluctuated up and down over the following two decades.
51. Knesset Research and Information Center, *Data on Crime in Arab Society in Israel* (Hebrew) (Itamar Milrad) (Jerusalem, November 12, 2014), 7–8.
52. Gal, *Haredim in Israeli Society*, 3. The fertility rate has in the past been as high as 7.62, though it has declined in recent years, and in 2009, it was 6.5 versus 2.9 for all Israeli Jews.
53. Eitan Regev, *Education and Employment in the Haredi Sector, Policy Paper No. 2013.06* (Jerusalem, June 2013), 19.

54. Guy Grimland, "Forget Switchboards, Haredi Women Can Be Great Engineers," *Ha'aretz*, February 20, 2011.
55. Regev, 20–21.
56. Shuki Handels, *Feelings of Discrimination Among Workers and Job Seekers and Diversity in Workplaces* (Hebrew) (Jerusalem, February 2013), 20–21.
57. Regev, 11.
58. Ibid.
59. Friedman, 76.
60. Roni Zinger, "How Much Does Slaughtering a Kosher Chicken Cost? 1.2 Billion Shekels a Year" (Hebrew), *Calcalist*, January 1, 2016.
61. Regev, 20–21.
62. Regev, 28.
63. National Insurance Institute, *Dimensions of Poverty and Social Gaps, 2013, 2014 and 2015* (Hebrew) (Jerusalem, December 2013, November 2014 and December 2015). In 1998, less than 5% ultra-Orthodox families with two breadwinners were under the poverty line, but by 2014 the rate had risen to 30%.
64. National Insurance Institute, Dimensions of Poverty and Social Gaps, 2014, 32.
65. Eitan Regev, "Making Ends Meet—Household Income, Expenditures and Saving in Israel," in *State of the Nation Report 2014* (Jerusalem, July 2014), 27. The figures are for 2011.
66. Ibid., 63.
67. Ibid., 65, 72–73.
68. Haim Zicherman and Lee Cahaner, *Modern Ultra-Orthodoxy: The Emerging Hi Middle Class in Israel* (Hebrew) (Jerusalem, 2012).
69. Reuven Gal, *Ultra-Orthodox Jews in Israel*, 16–17.

CHAPTER 12

Conclusions

The Israeli economy began assuming its modern shape some three decades ago from unpromising antecedents. Palestine and the Middle East were undeveloped socially and economically, and the Jews who began arriving in the late nineteenth century and would form the basis of the future state were committed to building a socialist economy based on agriculture. Palestinian and Arab resistance to the Zionist enterprise forced the *yishuv* and later the State of Israel to contend with war and terror as well as with political and economic isolation from its neighbors, all of which severely taxed its financial and human resources. Moreover, in its first years, Israel absorbed hundreds of thousands of impoverished and uneducated immigrants that strained its capacity to provide employment, housing, and infrastructure. These inauspicious beginnings, however, belied an ethos that would prove stronger than the objective conditions under which Israel was formed and developed. Zionism gave birth to a self-conscious culture determined to upend Jewish life as it was in Europe and to create a new society. The new man and woman would be, in the ideal, dedicated on a macroscale to building the state and on a microscale to his or her group—a collective settlement, an army unit, or a group of friends. Collectivism was a paramount value but so, seemingly contradictorily, was individual initiative. Candidness, thinking in terms of missions to be accomplished and problems to be solved, and a refusal to accept hierarchies or accepted ways were other critical elements of this culture. The fact that Zionism as a national movement and Israel as a state were at their heart an ideological enterprise—not the accident of

D. Rosenberg, *Israel's Technology Economy*, Middle East in Focus,
https://doi.org/10.1007/978-3-319-76654-6_12

geography and demography—perhaps explains how an ideal was able to overcome material conditions. At the outset, the province of a small, pioneering elite, these values came to permeate all of Israeli society and to form perhaps the single most important foundation for Israel's start-up economy.

That foundation, however, would not reveal itself until much later, when Israel's modern economy began to take shape in the middle 1980s. In the initial stages, it involved shedding the socialist doctrines that had guided it in the first years of the state and gradually dismantling the network of government-labor union companies that had once controlled two-thirds of the economy. The government simultaneously undertook a long process of deregulation and liberalization that continues to this day and in the early 2000s adopted a fiscal regime of low taxes and deficits. The arrival of hundreds of thousands of immigrants from the former Soviet Union in the 1990s strengthened Israel demographically and made an important contribution to its human capital. The 1990s also saw the rise of the Israeli high-tech industry, the result of both global developments (the rapid evolution of information technology and deregulation of telecommunications markets) and domestic factors (a strong human capital foundation and a culture that eagerly adopted the start-up ethos). All this occurred as the end of the Cold War and the first steps of the peace process with the Palestinians put an end to the relative isolation Israel had suffered, especially after 1967, paving the way for diplomatic and commercial relations with nearly all non-Arab world. Moreover, the existential threat to Israel's existence began to recede: Already in 1979, a peace agreement with Egypt had been reached, and in 1994, a pact with Jordan was signed. If Israel's other neighbors—Syria and Lebanon—have yet to formally end their state of war, their internal weakness and instability have prevented them from acting on it, especially in the wake of the Arab Spring in 2011. Iran and other, non-state actors still present a danger, but the sharp decline in Israel's defense spending relative to the size of the economy serves as a good barometer of how Israel and its leaders assess the threat. By the early 2000s, Israel had assumed its current economic structure: A knowledge sector based on a cluster of start-up companies side by side with a much larger economy that has largely failed to make use of the innovation in its midst.

All of this progress has been accompanied by setbacks. Israel has fought no less than four wars over the last 30 years and contended with two Palestinian intifadas. The peace process with the Palestinians was never brought to its final conclusion, and the ultimate status of the West Bank and Gaza Strip remains unresolved, a situation that weighs heavily on Israel, both as a source of tensions with the international community and for bitter conflict in its domestic politics. Despite this, Israel's political system has functioned well and is arguably more democratic and more inclusive of its variegated population than it has been any time in the past. As cataclysmic as it was, the assassination of Prime Minister Yitzhak Rabin in 1995 proved not to be an omen for a future of political extremism or violence. But few observers of the Israeli political scene would deny that the country has experienced to a degree the same bitter politics created by the rise of populism in the USA and Europe after 2008 financial crisis. To the extent that Israel has avoided the worst features of the angry politics of populism, in spite of the political traumas it has undergone, can be explained by the economy's ability to deliver sustained growth and jobs at a time when other developed countries have been struggling.

The knowledge economy that has arisen from these processes rests on a small foundation of start-up technology companies. Wholly dedicated to developing new products and services, few of these companies see themselves as businesses in the normal sense—enterprises that look to develop a sustainable business that will not only create new technology, but also market and/or manufacture it, develop follow-on products, and take on all the functions of a business aspiring to those goals. The great majority of the most successful of these start-ups are eventually sold, more often than not to a foreign multinational. Those that opt to remain independent and expand face a daunting challenge because the same qualities that enable Israelis to so successfully start-up companies and innovate are antithetical to the culture of big organizations. In America and Europe, the challenge of transitioning from a start-up culture to a big-company culture often entails the founders ceding control to managers with the skills required to lead a large organization; in Israel, however, there are relatively few managers who can lead big companies and they are more likely to encounter a deeper resistance from employees to a corporate of thinking and organization.

By itself, the start-up phenomenon is not a negative—it is a source of innovation, creates rewarding and well-paid jobs, attracts foreign investment, and in Israel's case has even become a political asset for the country in a world hungry for innovation. But it offers few knock-on effects for the wider economy; quite to the contrary, in a country where the educational system fails to provide quality schooling for the great majority, Israel's start-up industry monopolizes the best and the brightest. And because it has failed to develop bigger, sustainable enterprises, the start-up sector offers opportunities only to those with a narrow set of skills, nearly all of them in engineering related to information technology. It deprives the Israeli economy of the full benefits that innovation should be bringing because they are far more likely than not to accrue to the overseas companies that acquire the technology. Israel's global economic competitiveness does not rely wholly on its high-tech sector, but even outside of high tech it is overly dependent on a narrow base of companies.

Thus, while the economy has had excellent record over the last two decades of sustained growth and job creation, they are mostly in sectors that serve a protected domestic economy. The availability of relatively low-cost, under-trained labor—the inadvertent consequence of government policies—encourages this phenomenon. The result is that despite high levels of economic growth, Israeli GDP per capita is at the lower end of the developed world, labor productivity is poor, and it has struggled to reduce its high rate of poverty.

Despite Israel's evident successes, the overall tone of this book has been critical of Israel's economic performance. That is because in a dynamic and globally interconnected economy, there is no time to rest of one's laurels. Israel enjoyed a favorable confluence of economic and political factors that came together after 1985, but those achievements have come hand in hand with failures. Some of them, such as high rates of poverty, poor schools, and low rates of labor productivity, are already evident; others, like the challenging demographic outlook that will leave it with an older and less-educated population if current trends remain in place, will only manifest themselves in the years ahead. These failures, both the ones Israel faces now and the ones looming ahead, jeopardize the country's achievements of the last three decades. Analyzing them is as important, if not more important, than examining the factors that contributed to its successes. Moreover, the Israeli experience points to the limitations of a knowledge economy for creating a society where the wealth and opportunities it generates accrue to a large share of the population on a reasonably equitable basis. If Israel's start-up-focused knowledge economy makes these problems particularly acute, the fact is that even knowledge-based economies that rest on a wider base have suffered the same problems. Knowledge-based sectors by their nature bestow higher incomes and employment opportunities chiefly on those with the relevant education and skills. Those without them have fewer alternatives. The combination of automation, robotics and artificial intelligence, and the emigration of less-skilled work to less-developed economies threatens to narrow their possibilities even more.

Thus, the single biggest challenge facing Israel in the years ahead is expanding the base of its knowledge economy. Israel's most valuable education assets are its universities and the non-school educational resources, mainly the army. In the case of the first, Israel must substantially increase funding of higher education both to ensure access to as wide as possible of the population and to enhance their research and development capabilities. Israeli is producing fewer graduates as a percentage of the population, partly not only because the wave of Russian immigration raised the level artificially high but also because the growing Haredi community shuns secular education of any form and its share of the population is growing. Ensuring wide access to higher education won't by itself solve this problem, but it is a necessary component of the solution. Strong academic R&D is both critical for many segments of high tech, most notably biotechnology, but it is also important in order to retain the best academic talent and help create an internationally competitive education industry (as discussed below). The more serious challenge in education, however, is on the pre-tertiary level, where Israeli schools have failed at their task of providing basic skills. Funding isn't the central problem so much as it is institutional and pedagogical failure. That is a far more difficult problem to address than a money one, but if Israeli schools can't fulfill their task, the country's ability to widen its knowledge economy will be severe constrained. Israel is less in need of the innovators that make its start-up sector so effective than it is of people with the knowledge and skills to work effectively in more conventional knowledge-based business and public service.

The second challenge is integrating Israeli Arabs and Haredim into the workforce. To some degree, this is already happening, but it progressing too slowly given the demographic changes occurring in Israel that could see the two groups' combined increase their share of the population to close to half. If the educational and occupational parameters for Israeli Arabs and the ultra-Orthodox aren't improved, Israel will cease to have the critical mass of human capital a knowledge economy requires. Access to education is, of course, a critical factor, but it will take more than that because both populations right now exist as virtually separate societies. Israel's task therefore isn't simply to raise educational levels for them but to more fully integrate them into mainstream Israeli society. It's unrealistic and undesirable to imagine Israel as having three economies for three populations; yet, creating a single one, where all Israelis work and do business together and share common values and culture, requires comprehensive changes in social attitudes that are beyond the scope of policymakers to implement through the usual tools of laws, regulations, and funding. Vis-a-vis the Haredim, those changes will have to come from inside the community by surrendering the values that lie at the heart of their Society of Learners. This will only come if the economic pressures become too great to resist. Vis-a-vis Israeli Arabs, the work of social transformation will have to come mainly from Jewish Israelis abandoning the racist attitudes too many of them now hold. Certainly, a peaceful conclusion to Israel's conflict with the Palestinians would aid this process, but the economic imperative is too important for it to wait.

The third challenge is to improve the quality of government. In a geographically small and densely populated country like Israel, the government is inevitably going to play a major role in the allocation of resources like land, water, transportation, and the environment, to name a few. Moreover, a small embattled state like Israel requires a high level of the social solidarity to ensure that its citizens feel that the sacrifices and the benefits of being Israeli are shared on what is perceived as an equitable basis. All this requires a relatively big and activist government. But for the state to play its role effectively, it must be able to provide services effectively, which it currently does not by the standards of a developed economy. On the other hand, the government needs to also know when to pull back from its responsibilities in terms of rules and regulations in order to give the business sector the flexibility to operate efficiently. Here, too, the government is characterized by both a heavy-handed approach to rules-making while at the same time ineffectively enforcing many of those rules. Given Israeli culture's resistance to big organization and systems and the powerful role of unions in the civil service, there is a limit to how efficient the state sector can aspire to be. Nevertheless, policymakers should aspire to do better.

The fourth challenge relates to Israel's political and religious culture. The start-up ethos requires people to be able to think, speak, and act freely—qualities that can't flourish isolated in the workplace but must come within the context of a society that encourages and even cultivates them. These qualities have been the basis for Israel's start-up culture in equal measure to the country's science and engineering prowess. Short of a political revolution and a change of regime, it is difficult to determine whether a society is evolving from a freer to a less-free social and political environment. These processes occur slowly and not in a straight line, but in Israel's case, the trends are worrying. The last decade has seen the rise of a culture war between the country's older left-wing, Ashkenazi establishment and a right-wing of socially mobile Mizrahim, settlers, and Orthodox Jews. This war is often conducted in shrill voices, and it should be noted that it is taking place against a background of greater freedoms and inclusiveness than Israel has ever known, yet the nature of the war is that both sides seek to constrain freedoms—one in the name of national unity and the other in the name of inclusiveness and anti-racism. Meanwhile, the growing religiosity of Israelis presents a challenge to Israel's free-wheeling culture by its stress on piety. Again, this is not an area where policymakers by themselves can have much influence. Political leaders could have some effect by strongly expressing their support-free expression and pluralism, but most seem to show a preference for taking sides in the culture wars. Israelis will have to take it upon themselves to preserve their free and open culture.

These are the principal domestic challenges Israel faces, but there are also external threats that for now remain by Israeli standards muted but could grow—indeed, arise quiet suddenly—in the years ahead.

The threat of conventional war had receded, but it has by no means disappeared. Syria, the only adjacent country with the potential military power to threaten Israel, is preoccupied with the civil war that began in 2011. When it ends, it will take years for the country to recover. But if the war concludes with the Assad regime in power, Israel will likely be contending with a client state of the regime in Tehran. Under the circumstances, Syria might not pose a serious military challenge to Israel but together with the Lebanese militant group Hezbollah could engage Israel in a costly war of attrition. Looking at Israel's other neighbors, peace with Jordan and Egypt is firmly established and the two countries are ruled by regimes that are decades' old. But both countries face immense economic problems and popular discontent. The Egyptian political establishment was briefly toppled by a Muslim Brotherhood regime and a similar scenario could occur again, this time with more lasting results. In both countries, Islamism is a compelling ideology (perhaps the only compelling ideology in the Middle East) for a population with justifiable grievances against their rulers. It constitutes a fundamental threat to the region's established order and to Israel's security. Israel's ability to contend with the challenge is largely reactive; apart from security aid and, to a lesser extent, economic and technical assistance, it can do nothing to help address the core problems its neighbors face.

More than conventional war, however, the immediate security threat Israel must contend with is from the missile wars it has fought repeatedly in the 2000s. Israel has developed sophisticated anti-missile technology, most notably the Iron Dome system for tracking and taking down short-range rockets of the kind used by Hezbollah and Hamas in the Gaza Strip. The technology is a testament to the nexus between Israel's defense needs and the technology sector and proved itself during the 2014 Operation Protective Edge conflict. However, the missile threat is growing, and it is by no means clear that Israel can provide repeated technology upgrades against attacks involving many thousands of projectiles with bigger payloads. Israel's strategy of responding in kind to a mass attack is by its nature will result in a pyrrhic victory in a war that will almost certainly involve considerable human and material losses to both sides. Meanwhile, Israel is engaged in a Cold War with Iran which, if it one day erupts into a hot conflict, will be fought with missiles that will almost certainly inflict even more significant damage than Hezbollah and Hamas can ever hope to achieve. Finally, mass violence of the kind that characterized the Intifadas of the 1980s and early 2000s remains a real danger. A third intifada would not likely cause the same damage as a missile war, but a wave of terror attacks would take a huge toll on its economy and measures Israel is likely to take to contain it would almost certainly undermine its standing in the international community. That was certainly the lesson of the Second Intifada.

The external threat Israel faces is a political one as well. With the end of the Cold War and the onset of the peace process, Israel's diplomatic and trade relations have widened to encompass nearly all the rest of the non-Muslim world. More recently, Israel has forged especially strong and important ties with the emerging powers of China and India. Israel's high-technology prowess been a critical and often underappreciated factor in enhancing the country's global standing, not only by attracting foreign investment and multinational mergers and acquisitions but by acting as a basis for political relations as well. Technology is seen by governments as a critical part of industrial policy, a way of ensuring their economies remain competitive and for their industries to advance higher on the value chain. As a small country with limited industrial aspirations of its own, Israel is a ready source of intellectual property and an ideal partner for research and development collaboration. However, all of these achievements are at risk because of the Israeli–Palestinian dispute. Israel is more often than not perceived as the side preventing an agreement and as a violator of basic human rights, both due to the occupation and the way it conducts its frequent wars. Regardless of the validity of the accusations, they have unarguably influenced public opinion in the West and the way in which the Israeli–Palestinian conflict is framed in the media. The critique of Israel often fails to stop at Israeli policies and in its most extreme forms casts doubt of the country's right to exist. Since the end of the Second Intifada, this has expressed itself in its most extreme form in a call for businesses and governments to boycott Israel, divest holdings in Israeli companies, and impose sanctions. As of this writing, the impact has been largely restricted to nongovernmental organizations, some labor unions and academic organizations, student governments, and a handful of artists. Governments, international organizations, and businesses have not supported such calls, much less initiated them. As a result, the boycott movement has failed to inflict any economic damage, even if at times it can claim a small, high-profile victory, such as a performing artist's canceling an appearance. But there is a very real danger from Israel's point of view that the tide of public opinion will turn sufficiently to deter companies from doing business with Israel and/or leading governments to impose sanctions. Whether that happens will be decided as much by the decisions of Israeli policymakers as it is by the changing moods of Western public opinion.

Assuming Israel successfully contends with all these challenges, or at least is able to forestall them, what kind of knowledge economy should it aspire to? Even in the most optimistic scenario, one where the country solves the problem of widening its human capital base and creates a more business-friendly regulatory environment, a significant expansion of its start-up sector by itself would not address all the economy's needs for creating jobs and a more equitable distribution of wealth. To achieve these goals, below are five strategies Israel could employ to create business opportunities and employment in sectors where Israel has potential that has been only partially exploited or not exploited at all. In doing so, it would address the problems of poverty and inequality Israel now struggles with.

1. *Mittelstand* Companies: The German prototype—a small- or medium-sized enterprise with annual revenues up to 50 million euros, family-owned or at least closely held, and employing no more than 500 people—offers Israel both micro- and macro-economic benefits. In modern business, with frequent shifts in technology, regulations, and consumer tastes, the risk of a company failing is bigger than it has ever been in the past. For a small economy like Israel's, the risk is magnified if the company is a major employer and exporter whose downfall reverberates across the economy in the way that Finland suffered with the reversal of fortunes for Nokia after 2007. From a macroeconomic prospective, the development of a Mittelstand sector would obviate those risks. Another macroeconomic benefit that Israel could expect is due to the fact that big companies in the Israeli context must ipso facto be multinational and therefore diversify their sourcing, manufacturing, marketing, finance, and other operations across different countries. The result is that the direct benefit to the economy of giant multinationals in terms of employment is relatively less than a large number of SMEs. On a microeconomic level, Mittelstand companies' smaller scale and more intimate corporate culture is better suited to Israelis' resistance to large organizations. One important difference in the Israeli version of the Mittelstand would be to avoid the sectors favored by German companies (machine tools, auto parts, electrical equipment, and chemicals) and instead play to the country's strength in innovation and technology by developing companies that provide high value-added products in niche markets.
2. Advanced Manufacturing: This emerging sector is rapidly creating a radically new manufacturing environment that replaces machinery-driven mass production with a microprocessor-based factory floor. It encompasses a range of emerging technologies, including big data, computer networking, cloud commuting, 3D printing, low-cost electronic sensors, and new materials, that enable small and nimble manufacturing processes. In what is perhaps its most extreme manifestation, 3D printers create products directly from digital models using designs previously regarded as unfeasible, wrought from new and unconventional materials. Products can be produced in a short runs, even on a one-off basis, from a single printer that can be reconfigured to make a virtually limitless array of products. The advantages for Israel from advanced

manufacturing are obvious. Israel has the technology infrastructure to develop products and processes. The low-costs of acquiring the hardware and software, and the ability to repurpose them as the need arises, give smaller companies a gateway into manufacturing industries that were once only feasible for the biggest enterprises. The labor force required for advanced manufacturing is tiny relative to the needs of mass manufacturing, but workers do need to be relatively well-educated, able to fulfill multiple tasks, and solve problems. Their jobs don't involve physical labor and a highly coordinated operation based on a production line so much as management of the factory floor's hardware and software. Thus, advanced manufacturing meshes well with the characteristics of the Israeli labor force much the same way the Mittelstand business model does. Israel's principal shortcoming is its poor schools, which will need to produce better-trained graduates. As much as advanced manufacturing may rely more than traditional manufacturing on individual initiative and creative thinking, it also demands high levels of literacy and numeracy where Israeli workers now lag.

3. Industry Based on Natural Gas: Increasingly large reserves of natural gas have been discovered offshore Israel since the early 2000s. Production began on a small scale in 2004 and became a significant contributor to energy consumption after 2013 and will grow further still when Leviathan, the largest field discovered to date, goes on line about 2019. The government is determined to develop more fields in its Mediterranean waters. Official policy is to encourage natural gas exports in order to generate export receipts, enhance Israel's geopolitical stranding by becoming a major regional energy power, and to ensure the development of more fields as a buffer in the case existing ones are attacked or fail for technical reasons. But Israel should also be using gas as a catalyst to expand and create new industries.[1] As of 2015, about close to 80% of Israel's domestic gas consumption is used for generate electricity, with the rest being used directly by industry. This is a relatively low figure: In the USA, for instance, more than 40% of gas is used by industry. Israel's high-tech sector can make little use of the gas, but older industries, like paper, textiles, plastics, chemicals, and food, should be able to lower costs and improve their competitiveness by using gas a lower-cost energy source and/or as

an input. There is also potential for a methanol plant which would encourage secondary industries in chemicals and plastics. While the government estimated in 2012 that industrial use of gas would grow by 150% by 2020, the vast quantities of gas Israel has and its difficulties as of 2017 in developing export markets are good reasons for Israel to aspire to more. Industries based on natural gas would provide well-paid employment for the less educated and serve as a segue to the (hopeful) day Israel succeeds in upgrading its human capital base.

4. Higher Education: Israel counts a large number of top-ranked universities and justifiably basks in the glow of the country's reputation as a technology power, yet it has failed to fully capitalize on its system of higher education. The country's colleges and universities are correctly seen as sources for training the next generation and as centers of research and development. But if Israel adopts a policy of becoming a world teaching center, higher education could be source of employment and income as well. On a relatively modest scale, Israel has begun this process. Its universities have developed R&D ties with their Chinese counterparts, and the Technion Israel Institute of Technology is developing a joint science and engineering campus in New York City with Cornell University. Israel's Council for Higher Education adopted a five-year program starting in 2017 to more than double the number of overseas students studying at Israeli institutes of higher education to about 25,000.[2] This is a good start, although it should be noted that 60% of the projected number is slated for short-term programs rather than matriculating students. In the 2015/2016 school year, foreign students accounted for only 3.9% of the student population, including short-term students, and the CHE's plan would about double that. By comparison in Britain, 21.5% of the student population is from overseas. The figure is 21% in Australia, 13.3% in Canada, 13.1% in France, and 11.5% in Germany, suggesting that Israel has room to expand as the number of students globally studying abroad is growing, especially from emerging economies.[3] The number of jobs an education industry would create would not be large in absolute terms, but it would have considerable knock-on effects. It would enable the many Israeli academics working abroad to return home, enhancing Israel's R&D strengths in the process by expanding faculty numbers and funding and creating over the

years a cadre of graduates who would see Israeli companies as natural partners in future business relationships.

5. Creative Industries: Akin to education, Israel has considerable human resources in the media and creative arts that have not been fully exploited and could be put to use in creating employment and generating income in ways that were unimaginable 10 or 20 years ago. The global television market is a case in point and demonstrates how content is undergoing changes similar to what occurred in the telecommunication industry in the 1990s when the rise of the Internet combined with industry deregulation broke the lockhold of a limited number of monopoly operators and suppliers and threw open the market to high-tech start-ups in Israel and elsewhere. In global television, markets were controlled by a limited number of broadcasting networks that provided a relatively narrow range of programming designed to appeal to as broad an audience as possible. Production was costly and done by a limited number of content providers. The rise of the Internet to distribute content and digital technology to produce it has reduced costs and given rise to a plethora of new outlets such as Netflix, Amazon, Hulu, and YouTube for distributing it. Demand for content has grown exponentially, and broadcasters can address niche markets, creating new opportunities' for creators. In this context, Israel has emerged over the last decade as a prolific content provider, with hit series like "In Treatment" and "Homeland." Altogether, its three main production companies—Keshet International, Dori Media, and Armoza Formats—have developed original programming for some 100 shows to markets as diverse as the USA, Indonesia, Japan, Finland, and Brazil. Employment is very limited because these companies specialize in developing programming, not producing it, but Israel has the human capital to widen the business. It has the technology, including the Web analytics capabilities, to develop media businesses and markets, as well as a diverse population that can provide the cultural insights and range of languages needed to create international programming. The creative industry may even under the most optimistic scenarios create only a small number of jobs, but they would be rewarding and well paid and offer opportunities to Israelis without the skills and training required by the tech sector. In particular, Israeli Arabs could be a critical bridge to creating content for the Arabic-speaking world.

Importantly, these five industries and/or business models focus on Israel's export sector more than on the domestic economy. They apply the start-up model of small dynamic businesses and innovation to other sectors. Although in an ideal future, Israel would act to create a more dynamic and competitive business sector serving its home market together with better and more efficient regulation, these are areas that would require a revolution in government and Israeli culture that don't show any signs of being in the offing. Realistically, in the best case scenario, Israel might show some modest improvement in these metrics, but it seems too ambitious to expect anything more. Moreover, Israel's domestic market is small and relatively isolated, and the ability to foster true competition is limited; at some point, the benefits of competition created by multiple players in a single market are outweighed by the loss of economies of scale. It may therefore be Israel's fate to be a dynamic, competitive economy on the global level but perform less well in the service of its own citizens. For the hundreds of thousands of Israelis who took to the streets in the summer of 2011 demanding more from their government and from the economy, that would be a disappointing outcome, but in a world of often stark economic choices it is by no means intolerable. From the vantage point of 2017 to be an economy of enterprise and innovation of the kind Israel has evolved over the decades is a rare achievement and solid claim for a prosperous future.

Notes

1. The discussion on industrial applications for natural gas is based on: Gil Michael Bufman, Eyal Raz, and Noach Hager, *The Potential of Natural Gas in the Israeli Economy* (Tel Aviv, April 2014); and Yaniv Bar, *The Natural Gas Sector in Israel: An Economic Survey* (Tel Aviv, January 2017).
2. Israel Council of Higher Education, "Presentation of a New Multi-Year Program: 7 Billion Shekel Supplement to the Higher Education Budget" (Hebrew), undated.
3. *2016 Top Markets in Education, U.S. International Trade Administration*, 5. Estimate for Israel's overseas student population is author's calculations based on CHE data and includes short-term students. Not counting them, Israel's overseas student population is 2.3%.

Bibliography

2016 Top Markets in Education. Washington: U.S. International Trade Administration, undated. http://trade.gov/topmarkets/pdf/Education_Executive_Summary.pdf.

2017 Best Countries. *U.S. News & World Report*, undated. https://www.usnews.com/news/best-countries.

Abu Asba, Khaled, and Muhammad Abu Nasra, *The Palestinian Population in Israel at the Margins of the Israeli Economy*. Jerusalem: Israel Democracy Institute, November 8, 2012. http://en.idi.org.il/analysis/articles/the-palestinian-population-in-israel-at-the-margins-of-the-israeli-economy/.

Academic Ranking of World Universities. Shanghai Jiao Tong University, Shanghai, 2015. http://www.shanghairanking.com/ARWU2015.html.

Aharoni, Yair, *The Israeli Economy: Dreams and Realities*. London and New York: Routledge, 1991.

Ahmed, Murad, "Foreigners at the Helm of One in Five UK Tech Groups." *Financial Times*, March 28, 2016. https://www.ft.com/content/824b0836-f283-11e5-a609-e9f2438ee05b.

Ali, Nohad, *Representation of Arab Citizens in the Institutions of Higher Education in Israel*. Haifa-Jerusalem: Sikkuy—The Association for the Advancement of Civic Equality, November 2013. http://www.sikkuy.org.il/wp-content/uploads/2013/11/English_final-2014_representation_higher_education1.pdf.

Almog, Oz, *The Sabra: The Creation of the New Jew*. Berkeley, CA: University of California Press, 2000.

Amit, Hagai, "Mortgaging Their Future." *Ha'aretz*, September 29, 2012. http://www.haaretz.com/business/david-com-vs-goliath-inc/mortgaging-their-future.premium-1.467385.

D. Rosenberg, *Israel's Technology Economy*, Middle East in Focus,
https://doi.org/10.1007/978-3-319-76654-6

Amsterdamski, Shaul, "Full Boycott Report Revealed: A Complete Boycott of Israel Would Cost 40 Billion Shekels a Year" (Hebrew). *Calcalist*, July 7, 2015. http://www.calcalist.co.il/local/articles/0,7340,L-3663692,00.html.

Arlosoroff, Meirav, "Ignoring Arab Education Imperils Israel's Future." *Ha'aretz*, March 11, 2014. http://www.haaretz.com/israel-news/business/.premium-1.578944?date=1458564177402.

———, "Israel's Fat High-Tech Salaries Are Suddenly Going to Look a Lot Thinner." *Ha'aretz*, November 13, 2015.

———, "Does Studying the Talmud Make You Smarter? Data Dispels Old Myth." *Ha'aretz*, July 27, 2016.

———, and Sivan Klingbail, "Litzman: It's Absurd for Both Parents to Work—It's Enough That the Wife Works" (Hebrew). *TheMarker*, October 31, 2012. http://www.themarker.com/news/politics/1.1853906.

Asgeirsdottir, Berglind, "OECD Work on Knowledge and the Knowledge Economy." In Brian Kahin and Dominique Foray, eds., *Advancing Knowledge and the Knowledge Economy*. Cambridge, MA: The MIT Press, 2006.

Atad, Amnon, "The Reason for Economic Growth: Expansion at the Intel Plant" (Hebrew). *Calcalist*, August 19, 2012. http://www.calcalist.co.il/local/articles/0,7340,L-3580484,00.html.

Azoulay, Yuval, "Intel Exports in 2015: $4.1 Billion" (Hebrew). *Globes*, January 30, 2016. http://www.globes.co.il/news/article.aspx?did=1001099341.

———, "Ten Companies Exported More Than 51% of Israeli Merchandise" (Hebrew). *Globes*, March 20, 2016. http://www.globes.co.il/news/article.aspx?did=1001111409.

Azulay, Yoav, et al., *Facts and Numbers 2016* (Hebrew). Jerusalem: Israel Education Ministry, 2016. http://meyda.education.gov.il/files/MinhalCalcala/uvdot_venetunim2016.pdf.

Baller, Silja, Soumitra Dutta, and Bruno Lanvin (eds.), *The Global Information Technology Report 2016*. Geneva: World Economic Forum, 2016. http://www3.weforum.org/docs/GITR2016/GITR_2016_full%20report_final.pdf.

Bank of Israel, *Annual Reports, 2007, 2009, 2011, 2012, 2013 and 2014*. Jerusalem: Bank of Israel. The reports can be accessed at http://www.boi.org.il/en/NewsAndPublications/RegularPublications/Pages/Default.aspx.

———, *Recent Economic Developments*, No. 116 (July–December 2006). http://www.boi.org.il/develeng/develeng116/develeng.pdf.

———, *Recent Economic Developments*, No. 132 (September–December 2011). http://www.boi.org.il/develeng/develeng132/develeng.pdf.

———, "The Development of Education in Israel and Its Contribution to Long-Term Growth." *Recent Economic Developments*, No. 136 (April–September 2013). http://www.boi.org.il/en/NewsAndPublications/RegularPublications/Research%20Department%20Publications/RecentEconomicDevelopments/develop136e.pdf.

———, "Labor Productivity in Domestic-Market Oriented and Export Industries: An Analysis from an International Perspective." *Annual Report 2013*, April 2, 2014. http://www.boi.org.il/en/NewsAndPublications/RegularPublications/Research%20Department%20Publications/BankIsraelAnnualReport/Annual%20Report-2013/p2-2013e.pdf.

———, "The Connection Between the Quality of Education and Growth: Israel Compared to the World." *Recent Economic Developments*, No. 139 (October 2014–March 2015). http://www.boi.org.il/en/NewsAndPublications/RegularPublications/Research%20Department%20Publications/RecentEconomicDevelopments/red139e.pdf.

———, "Broader View of Selected Issues." *Fiscal Survey and Selected Research Analyses*, October 2016. http://www.boi.org.il/en/NewsAndPublications/RegularPublications/Research%20Department%20Publications/RecentEconomicDevelopments/red141e.pdf.

Bar, Ilanit, *The Weighting of the Middle Class and an Analysis of Changes in Recent Years* (Hebrew). Jerusalem: Knesset Research and Information Center, April 14, 2013. https://www.knesset.gov.il/mmm/data/pdf/m03157.pdf.

Bar, Yaniv, *The Natural Gas Sector in Israel: An Economic Survey*. Tel Aviv: Bank Leumi, January 2017. https://english.leumi.co.il/static-files/10/LeumiEnglish/Leumi_Review/NaturalGasinIsraelacc.pdf.

Barnea, Shlomit, Shurik Dryshpitz, and Ofer Kenig, "Government Stability—Working Paper" (Hebrew). Jerusalem: Israel Democracy Institute, undated. https://www.idi.org.il/media/6266/%D7%A0%D7%99%D7%99%D7%A8-%D7%A2%D7%91%D7%95%D7%93%D7%94_%D7%99%D7%A6%D7%99%D7%91%D7%95%D7%AA-%D7%9E%D7%9E%D7%A9%D7%9C%D7%95%D7%AA.pdf.

Bassok, Moti, "The Number of Avrechim Is Falling; 'The Path Is Still Long and Hard'" (Hebrew). *TheMarker*, June 21, 2016. http://www.themarker.com/news/1.2982229.

Belikoff, Michal, *Gaps Between Jews and Arabs in the Educational System—Physical Infrastructure* (Hebrew). Jerusalem: Sikkuy—The Association for the Advancement of Civil Equality, November 2014. http://www.sikkuy.org.il/wp-content/uploads/2014/12/%D7%A4%D7%A2%D7%A8%D7%99%D7%9D-%D7%91%D7%9E%D7%A2%D7%A8%D7%9B%D7%AA-%D7%94%D7%97%D7%99%D7%A0%D7%95%D7%9A-15.12.14.pdf.

Ben-David, Dan, "Brained Drained: CEPR Discussion Paper No. 6717," March 22, 2008. http://www.tau.ac.il/~danib/econ-rankings/BrainDrained.pdf.

———, "Update on the State of Israel's Universities and Its Researchers." In Dan Ben-David, ed., *State of the Nation Report 2013*. Jerusalem: Taub Center for Social Studies in Israel, November 2013. http://taubcenter.org.il/wp-content/files_mf/stateofnation_013eng8.pdf.

———, *The Shoresh Handbook on Israel's Society and Economy*. Tel Aviv: Shoresh Institution for Economic Research, 2015. http://shoresh.institute/index-publications.html.

———, and Haim Bleikh, "Poverty and Inequality over Time in Israel and the OECD." In Dan Ben-David, ed., *State of the Nation Report 2013*. Jerusalem: Taub Center for Social Studies in Israel, November 2013. http://taubcenter.org.il/wp-content/files_mf/stateofnation_013eng8.pdf.

Ben-Naim, Galit, and Alexey Belinsky, *Wage Divergence in Israel: Analysis of Wage Mobility Over the Last Decade* (Hebrew). Jerusalem: Joint Distribution Committee, June 11, 2013. http://www2.jdc.org.il/sites/default/files/tevet-doc-mobiliyut.pdf.

Ben-Porat, Amir, *State and Capitalism in Israel*. Westport, CT: Greenwood Press, 1993.

Bergman, Ronen, "How Israeli-Made Spyware Made It to the Arab World." *Ynet*, September 8, 2016. http://www.ynetnews.com/articles/0,7340,L-4851719,00.html.

Bior, Haim, "Wanted: 10,000 High Tech Workers at NIS 23,000 a Month" (Hebrew). *TheMarker*, May 30, 2012. http://www.themarker.com/career/1.1720131.

Botticini, Maristella, and Zvi Eckstein, "From Farmers to Merchants, Voluntary Conversions and Diaspora: A Human Capital Interpretation of Jewish History." *Journal of the European Economic Association*, 5(5): 885 (2007).

Bousso, Nimrod, and Zvi Zrahiya, "Buying a Home Will Cost an Average Israeli 146 Monthly Salaries." *Ha'aretz*, September 30, 2015. http://www.haaretz.com/israel-news/business/.premium-1.678119.

Brenitz, Dan, *Innovation and State: Political Choice and Strategies for Growth in Israel, Taiwan and Ireland*. New Haven, CT: Yale University Press, 2007.

Broude, Mark, Saadet Deger, and Somnath Sen, "Defense, Innovation and Development: The Case of Israel." *Journal of Innovations Economics and Management*, 2013/2, No. 12. https://www.cairn.info/revue-journal-of-innovation-economics-2013-2-page-37.htm.

B'Tselem: The Israeli Information Center for Human Rights in the Occupied Territories, "Attacks on Israeli Civilians by Palestinians." Jerusalem, July 24, 2014. http://www.btselem.org/israeli_civilians/qassam_missiles.

Bufman, Gil Michael, Eyal Raz, and Noach Hager, *The Potential of Natural Gas in the Israeli Economy*. Tel Aviv: Bank Leumi, April 2014. https://english.leumi.co.il/static-files/10/LeumiEnglish/PotentialofNaturalGasinIsraelApril2014.pdf?lang=en.

"Burden Sharing, and the Haredim." In Shlomo Fischer, ed., *Annual Assessment 2013–2014*, No. 10, Jerusalem: Jewish People Policy Institute, 2014. http://jppi.org.il/uploads/Burden_Sharing_and_the_Haredim.pdf.

Bystrov, Evengia, and Arnon Soffer, *Israel: Demography 2012–2030*. Haifa: Reuven Chaikan Chair for Geostrategy, May 2012.

Cahaner, Lee, Nikola Yozgof-Orbach, and Arnon Soffer, *Haredim in Israel: Space, Society, Community* (Hebrew). Haifa: Reuven Chaikan Chair for Geostrategy, November 2012.

Central Bureau of Statistics, "Survey of Educational and Welfare Services in Elementary and Post-elementary Schools—Hebrew and Arabic Education" (Hebrew). December 12, 2011. http://www.cbs.gov.il/webpub/pub/text_page.html?publ=84&CYear=2008&CMonth=1#100.

———, "International Aspects of Israeli Civilian Research and Development 2006–2009" (Hebrew). February 27, 2012. http://www.cbs.gov.il/www/hodaot2012n/12_12_049e.pdf.

———, "Innovation in the Business Sector, 2006–2008" (Hebrew). May 2012. http://www.cbs.gov.il/publications12/1463_innovation_06_08/pdf/h_print.pdf.

———, "ICT Sector Estimate for 2011: 1.5% Growth in GDP" (Hebrew). August 9, 2012. http://www.cbs.gov.il/reader/newhodaot/hodaa_template.html?hodaa=201229209.

———, "Departures and Returns in 2013 of Israelis Staying Abroad Continuously for a Year or More" (Hebrew). August 19, 2013. http://www.cbs.gov.il/reader/newhodaot/hodaa_template.html?hodaa=201501221.

———, "Educational Indicators for Training Professionals in Science and Technology in Israel" (Hebrew). January 14, 2014. http://www.cbs.gov.il/reader/newhodaot/hodaa_template.html?hodaa=201406009.

———, "Findings from the Business Innovation Survey 2010–2012" (Hebrew). July 10, 2014. http://www.cbs.gov.il/publications12/1463_innovation_06_08/pdf/intro_h.pdf.

———, Selected Data from the 2013 Social Survey About Employment, January 14, 2015. http://www.cbs.gov.il/reader/newhodaot/hodaa_template.html?hodaa=201519008.

———, *Defense Expenditures in Israel, 1950–2013*. Jerusalem, February 11, 2015. http://www.cbs.gov.il/webpub/pub/text_page.html?publ=7&CYear=2013&CMonth=1.

———, "International Women's Day 2015" (Hebrew). March 4, 2015. http://www.cbs.gov.il/reader/newhodaot/hodaa_template.html?hodaa=201511057.

———, "Immigration to Israel 2014" (Hebrew). May 19, 2015. http://www.cbs.gov.il/reader/newhodaot/hodaa_template.html?hodaa=201521129.

———, "ICT Sector Estimate for 2014: 4% Increase in GDP" (Hebrew). August 12, 2015. http://www.cbs.gov.il/reader/newhodaot/hodaa_template.html?hodaa=201529212.

———, "Family Day—Families and Households in Israel" (Hebrew). February 7, 2016. http://www.cbs.gov.il/reader/newhodaot/hodaa_template.html?hodaa=201611030.

———, *Statistical Abstract of Israel*, 2002, 2011, 2013 and 2016. http://www.cbs.gov.il/reader/shnatonenew_site.htm.

———, Divorces—Selected Data. http://www.cbs.gov.il/www/population/marrige_divorce/divorce_all.pdf.

———, General Government Defense Consumption, undated. http://www.cbs.gov.il/www/publications/defence/def_tab05.pdf.

———, Output, Gross Domestic Product and Exports of Information and Communications Technologies Industries, undated. http://www.cbs.gov.il/hodaot2016n/29_16_273t1.pdf.

———, *Social Survey 2008*, Table 33. http://www.cbs.gov.il/reader/cw_usr_view_SHTML?ID=927.

Chen, Yiyi, "China Is Destined to Intervene in the Conflict Between Israel and Palestine." *Quartz*, May 11, 2015. http://qz.com/401368/china-is-destined-to-intervene-in-the-conflict-between-israel-and-palestine.

Cingano, Federico, *Trends in Income Inequality and Its Impact on Economic Growth*. Paris: Organization for Economic Cooperation and Development, December 9, 2014. http://www.oecd-ilibrary.org/docserver/download/5jxrjncwxv6j.pdf?expires=1447853912&id=id&accname=guest&checksum=A4D7742FC46A616C8FF8067D6AE6599C.

Cochran, Gregory, Jason Hardy, and Henry Harpending, "Natural History of Ashkenazi Intelligence." *Journal of Biosocial Science*, 38(5): 659–693 (2006). http://harpending.humanevo.utah.edu/Documents/ashkiq.webpub.pdf.

Cohen, Gili, "IDF Statistics: One in Six Israeli Men Doesn't Complete Full Military Service." *Ha'aretz*, June 9, 2013. http://www.haaretz.com/1.528589.

Cohen, Ohad, "The Rise of Japanese-Israeli Economic Relations." *Times of Israel*, March 2, 2015. http://blogs.timesofisrael.com/the-rise-of-japanese-israeli-economic-relations/.

Cohen, Sarit, and Chang-Tai Hsieh, "Macroeconomic and Labor Market Impact of Russian Immigration in Israel." October 2000. http://www.biu.ac.il/soc/ec/wp/11-01/11-01.pdf.

Cohen, Tova, "Under Modi, Israel and India Forge Deeper Economic Ties." *Reuters*, November 19, 2014. http://www.reuters.com/article/2014/11/23/us-india-israel-ties-idUSKCN0J70L520141123.

Cohen, Uri, "The Hebrew University and Ethno-Class Stratification in the First Decade" (Hebrew). In Avi Bareli, Daniel Gutwein, and Tuvia Friling, eds., *Society and Economy in Israel: A Historical and Contemporary Perspective*. Jerusalem and Beersheva: Yad Yitzhak Ben-Zvi and Ben-Gurion University Press, 2005.

Cohen, Yinon, "Israeli-Born Emigrants: Size, Destinations and Selectivity." *International Journal of Comparative Sociology*, 52(1–2): 45–62 (2011). http://www.columbia.edu/~yc2444/ICS%202011%20emigration%20oecd.pdf.

Committee for Increasing Competitiveness in the Economy, Interim Report (Hebrew). Jerusalem, September 19, 2011. http://mof.gov.il/Lists/CompetitivenessCommittee_3/Attachments/1/MiddleReport.pdf.

Consolidated Edison Inc., 2014 Annual Report. file:///C:/Users/User/Downloads/ConEdison%202014%20Annual%20Report.pdf.

Coren, Ora, "Research: Productivity in Industrial High Tech in Israel Declined in the Last Decade Relative to the United States" (Hebrew). *TheMarker*, July 8, 2012.

———, "Today It's Clear That India and China Are Not an Immediate Threat to Israel" (Hebrew). *TheMarker*, July 11, 2012. http://www.themarker.com/technation/1.1753327.

———, "The Billion Dollar Question: What's an Exit Compared with Eternity" (Hebrew). *TheMarker*, August 7, 2012. http://www.themarker.com/magazine/1.1791038.

———, "Israeli Economist: Business with China Should Hinge on Long-Term Ventures." *Ha'aretz*, December 27, 2013. http://www.haaretz.com/business/.premium-1.565734.

Cornfeld, Ofer, and Oren Danieli, "The Origins of Income Inequality in Israel—Trends and Policy." *Israel Economic Review*, 12(2): 51–95 (2015). http://papers.ssrn.com/sol3/papers.cfm?abstract_id=2607192.

Corruption Perceptions Index. Berlin: Transparency International, 2011. http://cpi.transparency.org/cpi2011/results/.

Council for Higher Education, "Presentation of a New Multi-year Program: 7 Billion Shekel Supplement to the Higher Education Budget" (Hebrew), undated press release. http://che.org.il/%D7%A9%D7%A8-%D7%94%D7%97%D7%99%D7%A0%D7%95%D7%9A-%D7%95%D7%99%D7%95%D7%A8-%D7%94%D7%9E%D7%9C%D7%92-%D7%A0%D7%A4%D7%AA%D7%9C%D7%99-%D7%91%D7%A0%D7%98-%D7%99%D7%95%D7%A8-%D7%95%D7%AA%D7%AA/.

———, *The Higher Education System in Israel 2015* (Hebrew). Jerusalem, March 2015. http://che.org.il/?page_id=8383.

———, "Selected Data for the Opening of the Academic Year—2016–2017" (Hebrew). Jerusalem, October 10, 2016. http://che.org.il/?p=43805.

Dabla-Norris, Era, Kalpana Kochhar, Nujin Suphaphiphat, Frantisek Ricka, and Evridiki Tsounta, *Causes and Consequences of Income Inequality: A Global Perspective*. Washington: International Monetary Fund, June 2015. https://www.imf.org/external/pubs/ft/sdn/2015/sdn1513.pdf.

Dagan-Buzaglo, Noga, and Etty Konor-Attias, *Israel's Middle Class 1992–2010: Who Are We Talking About?* Tel Aviv: Adva Center, January 2013. http://adva.org/wp-content/uploads/2015/01/middle-class12.pdf.

Dahan, Momi, *Has the Melting Pot Succeeded in the Economic Field?* (Hebrew). Jerusalem: Israel Democracy Institute, September 2013. http://www.idi.org.il/media/2703168/%D7%9B%D7%95%D7%A8%20%D7%94%D7%99%D7%AA%D7%95%D7%9A%20-%20%D7%9E%D7%95%D7%9E%D7%99%20%D7%93%D7%94%D7%9F%20%D7%9C%D7%90%D7%99%D7%A0%D7%98%D7%A8%D7%A0%D7%98.pdf.

Dattel, Lior, "Data the Education Ministry Refuses to Publish: Haredi Student Achivement the Lowest in Israel" (Hebrew). *TheMarker*, November 8, 2012. http://www.themarker.com/career/1.1859732.

———, "TheMarker Investigation: In Secular Schools, Classroom Crowding Is Twice That in Religious Schools" (Hebrew). *TheMarker*, August 11, 2015. http://www.themarker.com/news/education/1.2704750.

Dovrat-Meseritz, Adi, and Sarit Menahem, "With 200 Million Shekels in the Bank, Rami Levi Wants to Be No. 1 Among the Retail Chains." June 22, 2012. http://www.themarker.com/markerweek/1.1737797.

Dror, Yuval, *Israelis in the Digital Age, 2012* (Hebrew). Rishon LeZion Israel: College of Management Academic Studies, May 2012. http://lib.ruppin.ac.il/multimedia_library/pdf/38374.pdf.

Duhigg, Charles, and Keith Bradsher, "How the U.S. Lost Out on iPhone Work." *The New York Times*, January 21, 2012. http://www.nytimes.com/2012/01/22/business/apple-america-and-a-squeezed-middle-class.html.

Dun & Bradstreet Israel, "The High Tech Industry—The State of Affairs." January 2017.

Dutta, Soumitra, and Beñat Bilbao-Osorio (eds.), *INSEAD, Global Information Technology Report, 2012: Living in a Hyperconnected World*. Geneva: INSEAD and the World Economic Forum, 2012. http://www3.weforum.org/docs/Global_IT_Report_2012.pdf.

———, *The Global Innovation Index 2012: Strong Innovations Linkages for Global Growth*. Fontainebleau, France: INSEAD and the World Intellectual Property Organization, 2012. http://www.globalinnovationindex.org/gii/index.html.

Eckstein, Zvi, and Daniel Tsiddon, "Macroeconomic Consequences of Terror: Theory and the Case of Israel." Prepared for Presentation at the Carnegie-Rochester Conference on Public Policy, Carnegie-Mellon University, November 21–22, 2003. http://www.tau.ac.il/~eckstein/pdf/Terror-dec9-03.pdf.

———, and Yoram Weiss, "The Integration of Immigrants from the Former Soviet Union in the Israeli Labor Market," May 2001. http://www.tau.ac.il-/~eckstein/pdf/Bruno_paper_english.pdf.

Education Ministry, *Initial Results of International Research—PISA 2012* (Hebrew). Jerusalem, December 3, 2012. http://cms.education.gov.il/EducationCMS/Units/Rama/MivchanimBenLeumiyim/PISA+2012.htm.

———, *Meitzav 5775 Part A, Achievement Tests* (Hebrew). Jerusalem: Ministry of Education, November 2015. http://meyda.education.gov.il/files/rama/Hesegim_Report_2015.pdf.

EGP Applied Economics, *The Venture Capital Industry in Israel: An Empirical Analysis of Economic Potential*. Tel Aviv: EGP, May 2009.

Eisenberg, Michael, "The Hummus Manifesto," undated. http://sixkidsandafulltimejob.blogspot.co.il/2010/07/hummus-manifesto-part-1.html.

Finance Ministry, *Israeli Labor Market-First Quarter 2015* (Hebrew). Jerusalem, undated. http://www.financeisrael.mof.gov.il/FinanceIsrael/docs/en/IsraeliLaborMarket_Q012015.pdf.

———, *Capital Markets, Insurance and Savings Division, Report of the Commissioner, 2015* (Hebrew), undated. http://mof.gov.il/hon/Pages/ReportsandPublications.aspx?LPF=Search&WebId=e6f49c89-90bd-40df-b09f-9ecae6009111&ListID=76773cbb-dfd0-4c3f-9960-da1db5f75cea&Item-ID=15&FieldID=MMDTypes_GxS_Invalid.

———, *Weekly Economic Survey* (Hebrew). February 14, 2016. http://www.mof.gov.il/chiefecon/economyandresearch/doclib/skiracalcalit_140216.pdf.

———, *Weekly Economic Survey* (Hebrew). August 21, 2016. http://www.mof.gov.il/chiefecon/economyandresearch/doundateclib/skiracalcalit_21082016.pdf.

Florida, Richard, and Charlotta Mellander, "The Geography of Inequality: Difference and Determinants of Wage and Income Inequality Across U.S. Metros." *Regional Studies*, 50(1) (2014). http://dx.doi.org/10.1080/00343404.2014.884275.

Freshfields Bruckhaus Deringer LLP, "Israel Passes Landmark Legislation Aimed at Breaking Up Conglomerates." December 2013. http://www.freshfields.com/uploadedFiles/SiteWide/Knowledge/Israel%20concentration%20law.pdf.

Freund, Oren, "Babylon Speaks the Tongue of the Markets." *Ha'aretz*, June 7, 2012. http://www.haaretz.com/business/babylon-speaks-the-tongue-of-the-markets.premium-1.434854.

Friedman, Amit, and Noam Zussman, "*Labor Force Quality in Israel*" (Hebrew). Jerusalem: Research Department, Bank of Israel, March 2008. http://www.boi.org.il/deptdata/mehkar/papers/dp0801h.pdf.

Friedman, Menachem, "*The Haredi (Ultra-Orthodox) Society—Sources, Trends and Processes*." Jerusalem: The Jerusalem Institute for Israel Studies, 1991. http://www.haredisociety.org/uploads/files/46498206618101313-haredcom.pdf.

Fuchs, Hadas, and Gilad Brand, *Education and Employment Trends Among Ethiopian Israelis.* Jerusalem: Taub Center for Social Policy Studies in Israel, June 2015. http://taubcenter.org.il/wp-content/files_mf/ethiopianeducationandemployment2015english.pdf.

Gal Reuven, *Ultra-Orthodox Jews in Israel: A Status Report, 2014* (Hebrew). Haifa: Samuel Neaman Institute, February 2015. http://www.kinneret.ac.il/Items/10678/%D7%94%D7%97%D7%A8%D7%93%D7%99%D7%9D%20%D7%91%D7%97%D7%91%D7%A8%D7%94%20%D7%94%D7%99%D7%A9%D7%A8%D7%90%D7%9C%D7%99%D7%AA.pdf.

———, *How to Bring Haredim into the Science and Technology Professions* (Hebrew). Haifa: Samuel Neaman Institute, June 2015. http://www.neaman.org.il/Neaman2011/Templates/ShowPage.asp?DBID=1&LNGID=2&TMID=581&FID=646&IID=10800.

Gao, Xiaohui, Jay R. Ritter, and Zhongyan Zhu, "Where Have All the IPOs Gone?" Cambridge, MA: Harvard Business School, April 3, 2012. http://www.hbs.edu/units/finance/pdf/Where%20Have_April_3_2012.pdf.

Getz, Daphne, et al., *Indices of Science, Technology and Innovation in Israel: An International Comparison* (Hebrew). Haifa: Samuel Neaman Institute for National Policy Research, 2013. file:///C:/Users/User/Documents/OLD%20COMPUTER/DAVIDS%20COMPUTER/MY%20DOCUMENTS/TechBook/Manuscript/6-382.pdf.

Goodtree, David, *The Massachusetts-Israel Economic Relationship.* Boston: New England-Israel Business Council, December 2013. http://neibc.org/boston-israel-study/.

Grimland, Guy, "Forget Switchboards, Haredi Women Can Be Great Engineers." *Ha'aretz*, February 20, 2011. http://www.haaretz.com/israel-news/business/forget-switchboards-haredi-women-can-be-great-engineers-1.34449.

Handels, Shuki, *Feelings of Discrimination Among Workers and Job Seekers and Diversity in Workplaces* (Hebrew). Jerusalem: Ministry of Industry, Trade and Employment, February 2013. http://www.economy.gov.il/Research/Documents/DiscriminationFeelings2013.pdf.

Harari, Keren Tsuriel, "The Middle Class' Secret Oxygen Supply: We Have an Account with the Parents" (Hebrew). *Calcalist*, June 27, 2013. http://www.calcalist.co.il/local/articles/0,7340,L-3606096,00.html.

Hays, Michael D., "The Training of Military Pilots: Men, Machines and Methods." Thesis Presented to the Faculty of the School of Advanced Airpower Studies, Air University Maxwell Air Force Base, Alabama, June 2002. https://www.researchgate.net/publication/235095640_The_Training_of_Military_Pilots_Men_Machines_and_Methods.

Hermann, Tamar, Nir Atmor, Ella Heller, and Yuval Lebel, *The Israel Democracy Index 2012.* Jerusalem: The Israel Democracy Institute, 2012. https://en.idi.org.il/publications/8769.

———, Ella Heller, Chanan Cohen, and Dana Bublil, *The Israel Democracy Index 2015*. Jerusalem: The Israel Democracy Institute, 2015. https://en.idi.org.il/publications/4076.

Hilou, Nadia, and Idan Haim, "Civilian Service in Israel's Arab Society." *Strategic Assessment*, 17(1): 59 (April 2014). http://www.inss.org.il/uploadImages/systemFiles/Hilou%20and%20Haim_adkan17_1ENG4.pdf.

Hirschauge, Orr, "Profile of Israeli High Tech: A Young Male, Veteran of an IDF Combat or Tech Unit and a University Graduate" (Hebrew). *TheMarker*, March 11, 2013. http://www.themarker.com/technation/1.1959126.

———, "Israeli IT People Are Being Left Behind: Advances in Technology But Not in Business" (Hebrew). *TheMarker*, May 7, 2013. http://www.themarker.com/technation/1.2013601.

Hoffman, Tzahi, "All the Numbers Behind the 261,000 High Tech Workers" (Hebrew). *Globes*, July 17, 2011. http://www.globes.co.il/news/article.aspx?did=1000665382.

Hoyle, Craig, et al., "World Air Forces: 2016." *Flight International*, December 3, 2015. https://d3fod5fkpt74ph.cloudfront.net/f5975bad82fe4e9e8d7f2e21e2e6f39e/3395ff5ac4d54a05bf6baa0c63895356.pdf.

Interministerial Committee Examining Competition and Prices in the Food and Consumer Product Sector (Kedmi Committee), Interim Report (Hebrew). Jerusalem, November 11, 2011. http://www.moit.gov.il/NR/rdonlyres/540E4C46-72F9-4542-A9B2-B759320B756F/0/doh_mazon.pdf.

Israel Advanced Technology Industries, *Israel's Life Science Industries 2015*, Herzilya Pituach: IATA, May 10, 2015. http://www.iati.co.il/files/files/Life%20Sciences%20Industry%202015.pdf.

Israel Electric Corp., *2014 Annual Report*. https://www.iec.co.il/EN/IR/Documents/FinancialReportsDecember2014.pdf.

Israel Export Institute, Economics Department, "Developments and Trends in Israeli Exports." Tel Aviv: Israel Export Institute, March 2016. http://www.export.gov.il/uploadfiles/04_2016/ieicidevelopmentsand%20trendsin%20israelexports15summary.pdf?redirect=no.

Israel Venture Capital Research Center, *IVC 2011 Yearbook*. Tel Aviv: IVC Research Center, 2011.

———, "2014 Israeli High Tech Exits Reach $6.94 Billion." January 6, 2015. http://www.ivc-online.com/Portals/0/RC/Exits/IVC-Meitar_Exits_2014_PR_Final.pdf.

———, "2015 Israeli High Tech Exits Hit $9.02 Billion—Up 16% from 2014 Proceeds." January 11, 2015. http://www.ivc-online.com/Portals/0/RC/Exits/2015/IVC-Meitar_Exits_2015_PR-final.pdf.

———, "Israeli Startup Success Report 1999–2014" (Summary Statement). January 28, 2015. http://www.ivc-online.com/Portals/0/RC/Startup%20Success/Startup%20Report%201999-2014_PR_Final.pdf.

———, *IVC 2016 Yearbook*. Tel Aviv: IVC Research Center, 2016. http://www.ivc-online.com/ivc-16/html/index.html#p.

———, "Israeli High Tech Exits at $10 Billion in 2016: 93 M&A Deals, 8 Buyouts and 3 IPOs." January 3, 2017. http://www.ivc-online.com/Portals/0/RC/Exits/IVC-Meitar_Exits_2016_PR-Final.pdf.

———, and Reversexit, "Only Four of 100 Start-Ups Succeed; Only Four of 500 Are Successful Growing Independently." January 28, 2015. http://www.ivc-online.com/Portals/0/RC/Startup%20Success/Startup%20Report%201999-2014_PR_Final.pdf.

Israeli Executives and Founders Forum. http://www.ieff.us/. http://ieff.us/map/.

Johnson, David E., et al., *Preparing and Training for the Full Spectrum of Military Challenges: Insights from the Experiences of China, France, the United Kingdom, India, and Israel*. Santa Monica, CA: Rand Corporation, 2009. http://www.rand.org/content/dam/rand/pubs/monographs/2009/RAND_MG836.pdf.

Keidar, Roy, "Israel: The Rise of the Fintech Hybrid." *Financier Worldwide*, July 2016. https://www.financierworldwide.com/israel-the-rise-of-the-fintech-hybrid/#.WPcsJmmGPIV.

Kesmodel, David, and Owen Fletcher, "Hummus Is Conquering America." *The Wall Street Journal*, April 30, 2013. http://www.wsj.com/articles/SB10001424127887323798104578453174022015956.

Kimhi, Ayal, "Trends in the Labor Market: Gaps in Employment Rates and Wages." *State of the Nation Report 2011–12*. Jerusalem: Taub Center for Social Policy Studies in Israel, undated. http://taubcenter.org.il/wp-content/files_mf/h2012.07kimhi.pdf.

King, Ian, "Intel's Israelis Make Chip to Rescue Company from Profit Plunge." *Bloomberg News*, March 27, 2007.

Knesset Research and Information Center, *Data on Crime in Arab Society in Israel* (Yaniv Ronen) (Hebrew). Jerusalem, February 23, 2010. http://www.qsm.ac.il/userfiles/elwaldi%20wlosra/mwad_nazareia/%D7%A0%D7%AA%D7%95%D7%A0%D7%99%D7%9D%20%D7%A2%D7%9C%20%D7%A4%D7%A9%D7%99%D7%A2%D7%94%20%D7%91%D7%97%D7%91%D7%A8%D7%94%20%D7%94%D7%A2%D7%A8%D7%91%D7%99%D7%AA.pdf.

———, *Main Influences on Immigration from the Commonwealth of States After 1990 on the Israeli Economy* (Ilanit Bar) (Hebrew). April 24, 2012. https://www.knesset.gov.il/mmm/data/pdf/m03149.pdf.

———, *Data on the Cost of Living in Israel in Comparison to Developed Countries: An Update* (Itamar Milrad) (Hebrew). November 12, 2014. https://www.knesset.gov.il/mmm/data/pdf/m03445.pdf.

———, *Study of the Possible Impact of an Economic Boycott of Israel* (Eyal Kauffman) (Hebrew). December 31, 2014. http://www.knesset.gov.il/mmm/data/pdf/m03501.pdf.

———, *Follow-Up on the Implementation of the Law for Promoting Competition in the Food Sector, 2014* (Itamar Milrad) (Hebrew). June 8, 2015. http://www.knesset.gov.il/mmm/data/pdf/m03549.pdf.

———, *Data on Haredi Employment* (Neta Moshe) (Hebrew). March 14, 2016. https://www.knesset.gov.il/mmm/data/pdf/m03737.pdf.

Knowledge@Wharton, "Why Tech Transfer Is Key to Stronger China-Israel Ties." August 13, 2015. https://knowledge.wharton.upenn.edu/article/why-tech-transfer-is-key-to-stronger-china-israel-ties/.

Kochhar, Rakesh, *Middle Class Fortunes in Western Europe*. Washington: Pew Research Center, April 24, 2017. http://www.pewglobal.org/2017/04/24/middle-class-fortunes-in-western-europe/.

Kop, Yaacov, "The Israeli Economy in the Fifth Decade Following D. Patinkin's the Israeli Economy in the First Decade." In Avi Bareli, Daniel Gutwein, and Tuvia Friling, eds., *Society and Economy in Israel: A Historical and Contemporary Perspective* (Hebrew). Jerusalem and Beersheva: Yad Yitzhak Ben-Zvi and Ben-Gurion University Press, 2005.

Kosenko, Konstantin, "Evolution of Private Business Groups in Israel: Their Impact at the Level of the Firm and the Economy." *Israel Economic Review*, 5(2): 55–93 (2007).

Krakowski, Moshe, "Worldview Construction and Identify Formation in Ultra-Orthodox Jewish Elementary Schools." *Diaspora, Indigenous and Minority Education: Studies of Migration, Integration, Equity and Cultural Survival*, 7 (2013, December 20, 2012). http://www.tandfonline.com/doi/abs/10.1080/15595692.2012.742052?journalCode=hdim20.

Landau, David, *Piety and Power: The World of Jewish Fundamentalism*. New York: Hill & Wang, 1993.

Laquer, Walter, *History of Zionism*. New York: Holt, Rinehart and Winston, 1972.

Leon, Nissim, *Moshe Kahlon and the Politics of the Mizrahi Middle Class, Research Paper 14*. College Park, MA: The Joseph and Alma Gildenhorn Institute for Israel Studies, January 2015. http://israelstudies.umd.edu/Nissim%20Leon%20Research%20Paper%20-%20January%202015.pdf.

Levine, Hagai (ed.), *The Haredi Sector in Israel: Empowerment Through Employment Integration* (Hebrew). Jerusalem: National Economic Council, Prime Minister's Office, March 2009. http://www.pmo.gov.il/BranchesAndUnits/eco/Documents/UltraOrtodox.pdf.

Levinson, Charles, and Adam Entous, "Israel's Iron Dome Defense Battled to Get off Ground." *The Wall Street Journal*, November 26, 2012. http://www.wsj.com/articles/SB10001424127887324712504578136931078468210.

Liebovitz-Dar, Sara, "In the Heart a Wall: The Big Battle of the Educational System over the Future of the Country" (Hebrew). *Ma'ariv*, May, 16, 2015. http://www.maariv.co.il/news/israel/Article-477157.

Liyan, Chen, and Andrea Murphy, "The World's Biggest Public Companies." *Forbes*, May 6, 2015. http://www.forbes.com/global2000/#country:Israel.

Menipaz, Ehud, Yoash Avrahami, Harel Yedidsion, and Miri Lerner, *Global Entrepreneurship Monitor 2010 Israel National Entrepreneurship Report.* Beersheva, Israel: Ben-Gurion University, 2010. http://www.gemconsortium.org/report#national-reports.

Menipaz, Ehud, Yoash Avrahami, and Miri Lerner, *National Report for Israel by the Global Entrepreneurship Monitor 2013 Global Entrepreneurship Monitor.* Beersheva, Israel: Ben-Gurion University, undated. file:///C:/Users/User/Documents/OLD%20COMPUTER/DAVIDS%20COMPUTER/MY%20DOCUMENTS/TechBook/Enetrepreneurism%20notes/1419513868GEM_2013_ENG_Israel_Report.pdf.

Merk, Olaf, and Thai Thanh Dang, *Efficiency of World Ports in Container and Bulk Cargo (Oil, Coal, Ores and Grain).* Paris: OECD Publishing, 2012. http://www.oecd-ilibrary.org/docserver/download/5k92vgw39zs2-en.pdf?expires=1490778861&id=id&accname=guest&checksum=7F8AD5775F3959949ECB4FC2D0E83A43.

Miaari, Sami, Nabil Khattab, and Ola Nabwani, *Occupational Mismatch Among College-Educated Arabs in Israel* (Hebrew). Jerusalem: Israel Democracy Institute, November 5, 2012. http://en.idi.org.il/analysis/articles/occupational-mismatch-among-college-educated-arabs-in-israel/.

Milgrom, Maor, *Economic Segregation in Israel, 1983–2008* (Hebrew). Tel Aviv: Institute for Structural Reforms, February 2015. http://reformsinstitute.org/cities/#/start/.

Ministry of Industry, Trade and Employment, *Invest in Israel, Statistical Profile.* Jerusalem, August 2011. http://www.investinisrael.gov.il/NR/rdonlyres/7BF899B2-2A8B-4B9D-8771-F9770EFC3C1E/0/ForeignRnDCentersinIsrael.pdf.

Mizroch, Amir and Orr Hirschauge, "Foreign Entrepreneurs Face Israeli Immigration Hurdles." *The Wall Street Journal*, September 23, 2014. http://www.wsj.com/articles/foreign-entrepreneurs-face-israeli-immigration-hurdles-1411490017.

———, "In Israel, Startups Race Toward Exit," December 22, 2014. http://blogs.wsj.com/digits/2014/12/22/in-israel-startups-race-toward-exit.

Myers-JDC-Brookdale Institute, *The Arab Population in Israel: Facts and Figures 2012.* Jerusalem, March 2012. http://brookdale.jdc.org.il/_Uploads/dbsAttachedFiles/Myers-JDC-Brookdale-Facts-and-Figures-on-Arab-Israelis-March-2012.pdf.

National Economic Council, *The Skilled Manpower Shortage in High Technology: Recommendations of the Inter-Ministerial Committee* (Hebrew). Jerusalem: Prime Minister's Office, July 2012. http://www.pmo.gov.il/Lists/FAQkalkala/Attachments/21/tech.pdf.

National Insurance Institute, *Annual Report 2008 and 2014* (Hebrew). Jerusalem: NII, June 2009 and September 2015. http://www.btl.gov.il/Publications/Skira_shnatit/skira-2008/Documents/skira-2008.pdf. http://www.btl.gov.il/Publications/Skira_shnatit/2014/Documents/skira%202014_hebrew.pdf.

———, *Dimensions of Poverty and Social Gaps, 2012, 2013 and 2014* (Miri Endeweld, et. al., eds.) (Hebrew). Jerusalem: NII, December 2013, November 2014 and December 2015. http://www.btl.gov.il/Publications/oni_report/Documents/DohOni2012.pdf. http://www.btl.gov.il/Publications/oni_report/Documents/oni2013.pdf. https://www.btl.gov.il/Publications/oni_report/Documents/oni2014.pdf.

Nisanov, Zoya, *The Middle Class in Israel.* Jerusalem: Taub Center for Social Policy Studies in Israel, July 2014. http://taubcenter.org.il/wp-content/files_mf/themiddleclassinisrael.pdf.

Office of the Chief Scientist (Israel), *Innovation Report* (Hebrew). Jerusalem: Israel Economy Ministry, April 13, 2015. http://www.economy.gov.il/Publications/PressReleases/Pages/InnovationReport.aspx.

Orbach, Meir, "Economy Ministry Approves Grating Visa for High Tech Entrepreneurs from Overseas" (Hebrew). *Calcalist*, October 20, 2015. http://www.calcalist.co.il/local/articles/0,7340,L-3671416,00.html.

Organization for Economic Cooperation and Development, Better Life Index. http://www.oecdbetterlifeindex.org/countries/israel/.

———, *Economic Survey for Israel, 2011.* Paris: OECD Publishing, 2011. http://www.oecd.org/dataoecd/38/45/49165941.pdf.

———, *Divided We Stand: Why Inequality Keeps Rising.* Paris: OECD Publishing, December 2011. http://www.oecd.org/els/socialpoliciesand-data/dividedwestandwhyinequalitykeepsrising.htm#NEWS.

———, *Education GPS, Student Performance (PISA 2012).* http://gpseducation.oecd.org/CountryProfile?primaryCountry=ISR&treshold=10&topic=PI.

———, *Economic Survey for Israel, 2013.* Paris: OECD Publishing, 2013. http://www.keepeek.com/Digital-Asset-Management/oecd/economics/oecd-economic-surveys-israel-2013_eco_surveys-isr-2013-en#page14.

———, *Education GPS, Israel, Teachers and Teaching Conditions*, 2013. http://gpseducation.oecd.org/CountryProfile?primaryCountry=ISR&treshold=10&topic=TA.

———, *PISA 2012 Results: Excellence Through Equity: Giving Every Student the Chance to Succeed Volume II.* 2013. https://www.oecd.org/pisa/keyfindings/pisa-2012-results-volume-II.pdf.

———, *Science, Technology and Industry Scoreboard 2013*. October 23, 2013. http://www.oecd-ilibrary.org/docserver/download/9213051e.pdf?-expires=1437574039&id=id&accname=guest&checksum=F681BB-8429B22A69E976AEFE3F88D7C3.

———, *Education at a Glance 2014: OECD Indicators*. Paris: OECD Publishing, October 2014. http://www.oecd.org/edu/Education-at-a-Glance-2014.pdf.

———, *Economic Policy Reforms 2015: Going for Growth*. Paris: OECD Publishing, February 9, 2015. http://www.keepeek.com/Digital-Asset-Management/oecd/economics/economic-policy-reforms-2015_growth-2015-en#page225.

———, *The 2013 Update of the OECD's Database on Product Market Regulation*. March 31, 2015. http://www.oecd-ilibrary.org/docserver/download/5js-3f5d3n2vl-en.pdf?expires=1484477927&id=id&accname=guest&checksum=FE34EA49EC91BFF40D5837D1F6BABF0D.

———, *Entrepreneurship at a Glance 2013*. August 5, 2015. http://www.oecd-ilibrary.org/docserver/download/3015021e.pdf?expires=1472633933&id=id&accname=guest&checksum=C24F6F49B-4CDA47F063AB1B471D1DCAC.

———, *How's Life? 2015: Measuring Well-Being*. October 13, 2015. http://dx.doi.org/10.1787/how_life-2015-en.

———, *Science, Technology and Industry Scoreboard 2015*. October 19, 2015. http://www.oecd-ilibrary.org/docserver/download/9215031e.pdf?expires=1480851800&id=id&accname=guest&checksum=49FB8F8465FF-4FAC90CE4E01655292C2.

———, *Taxing Wages 2014–2015 and Recent Trends*. 2016. http://www.oecd-ilibrary.org/taxation/taxing-wages-2015_tax_wages-2015-en.

———, *Economic Surveys Israel 2016*. Paris: OECD Publishing, January 2016. https://www.oecd.org/eco/surveys/Israel-Overview-OECD-Economic-Survey-2016.pdf.

———, Fertility Rates (database). https://data.oecd.org/pop/fertility-rates.htm.

———, Income Inequality (database). https://data.oecd.org/inequality/income-inequality.htm.

———, Young Population (database). https://data.oecd.org/pop/young-population.htm#indicator-chart.

Orpaz, Inbal, "The IDF's Apps Store" (Hebrew). *TheMarker*, May 18, 2015. http://www.themarker.com/technation/1.2638591.

Orpaz, Inbal, and Orr Hirschauge, "Who's at the Forefront of Israeli Innovation?" *Ha'aretz*, August 8, 2013. http://www.haaretz.com/business/.premium-1.540333.

———, "Game over in Download Valley?" *Ha'aretz*, August 23, 2013. http://www.haaretz.com/business/.premium-1.542896.

———, "Israeli Mafia Thrives in Silicon Valley but Can't Escape High Cost of Living." *Ha'aretz*, January 29, 2014. http://www.haaretz.com/israel-news/business/.premium-1.571087.

———, "The Secret to High-Tech Success? This Elite Israeli Army Unit." *Ha'aretz*, April 18, 2014. http://www.haaretz.com/israel-news/.premium-1.585863.

Paltiel, Ari, Michel Sepulchre, Irene Kornilenko, and Martin Maldonado, *Long-Range Population Projections for Israel: 2009–2059* (Hebrew). Jerusalem: Israel Central Bureau of Statistics, March 21, 2012. http://www.cbs.gov.il/publications/tec27.pdf.

Peretz, Gad, "Consulting Firm: 'Cellular Rates in Israel Likely to Rise'" (Hebrew). *Globes*, August 30, 2015. The Consulting Firm Cited Was Prometheus. http://www.globes.co.il/news/article.aspx?did=1001065203.

Plessner, Yakir, *The Political Economy of Israel: From Ideology to Stagnation*. Albany, NY: State University of New York Press, 1994.

Property Index Overview of European Residential Markets, fourth edition. Deloitte Touche Tohmatsu, June 2015. http://www2.deloitte.com/content/dam/Deloitte/be/Documents/realestate/be_fa_Deloitte-Property-index-2015.pdf.

PwC Israel, *The PwC Israel 2014 Hi-Tech Exit Report*, undated. http://www.pwc.com/il/en/technology/pwc_israel_exits_report_2014_en.pdf.

PwC-National Venture Capital Association, MoneyTree Report, Historical Trend Data. http://www.pwcmoneytree.com/HistoricTrends/CustomQueryHistoricTrend.

Reed, John, "Israeli Entrepreneurs and Immigration: 'Start-Up Nation' Stifled." *Financial Times*, November 11, 2014. http://www.ft.com/intl/cms/s/0/59bf7330-68f9-11e4-b389-00144feabdc0.html#axzz3dhZP0eOL.

———, "Israeli Tech Start-Ups Find Open Lane in New Automotive World." *Financial Times*, September 13, 2016. https://www.ft.com/content/985ef9e2-76a0-11e6-bf48-b372cdb1043a.

Regev, Eitan, *Education and Employment in the Haredi Sector, Policy Paper No. 2013.06*. Jerusalem: Taub Center for Social Policy Studies in Israel, June 2013. http://taubcenter.org.il/wp-content/files_mf/edu_emp_haredimtaubcenter.org.il_tauborgilwp_wpcontent_uploads_e2013.06haredim3.pdf.

———, "Making Ends Meet—Household Income Expenditures and Saving in Israel." *State of the Nation Report 2014*. Jerusalem: Taub Center for Social Policy Studies in Israel, July 2014. http://taubcenter.org.il/wp-content/files_mf/e2014.07makingendsmeet67.pdf.

Report of the Committee for Socio-Economic Change (Trajtenberg committee) (Hebrew). Jerusalem, October 2011. http://hidavrut.gov.il/sites/default/files/%20%D7%A1%D7%95%D7%A4%D7%99.pdf.

Report of the Reform Committee for Improving Human Capital Management in the Public Sector (Hebrew). Jerusalem, June 25, 2013. http://www.csc.gov.il/DataBases/ArticlesAndPublications/Documents/reformReport2013.pdf.

Report of the War on Poverty Committee, Part I (Hebrew). Jerusalem, June 23, 2014. http://molsa.gov.il/focus/documents/%D7%93%D7%95%D7%97%20%D7%94%D7%A2%D7%95%D7%A0%D7%99%2022.6.14.pdf.

Rooney, Ben, "How Entrepreneurial Is Europe?" *The Wall Street Journal*, November 11, 2013. http://blogs.wsj.com/tech-europe/2013/11/11/how-entrepreneurial-is-europe/.

Rosenberg, David, *Cloning Silicon Valley: The Next Generation of High-Tech Hotspots*. London: Pearson Education (Reuters Books), 2001.

———, "Chinese Investment in Israeli Tech Soaring." *Ha'aretz*, June 3, 2015. http://www.haaretz.com/business/.premium-1.659301.

Rosenberg, Oz, "Some 450,000 Israelis March at Massive 'March of the Million' Rallies Across Country." *Ha'aretz*, September 3, 2011. http://www.haaretz.com/news/national/some-450-000-israelis-march-at-massive-march-of-the-million-rallies-across-country-1.382366.

Rowen, Henry S., "Serendipity of Strategy: How Technology and Markets Came to Favor Silicon Valley." In Chong-Moon Lee, William F. Miller, Marguerite Gong Hancock, and Henry S. Rowen (eds.), *The Silicon Valley Edge: A Habitat for Innovation and Entrepreneurship*. Stanford, CA: Stanford University Press, 2000.

Rozovsky, Liza, "Educated Immigrants from the Soviet Union Boost the Job Marker—Their Children Are Under the Average." *Ha'aretz*, December 7, 2015. http://www.haaretz.co.il/news/education/.premium-1.2792762?=&ts=_1478609228844http://www.haaretz.com/israel-news/1.733691#article-comments.

Rubin, Eliran, "As Vehicles Go High-Tech, Israel's in the Driver's Seat." *Ha'aretz*, October 7, 2016. http://www.haaretz.com/israel-news/business/1.746264.

———, "Israel's Drone Industry Angry over Draconian Defense Export Restrictions." *Ha'aretz*, November 30, 2016.

Rubinstein, Yona, *School of Economics Policy Paper: Should Additional Seaports Be Developed and if so Under Government or Private Ownership?* (Hebrew). Herzliya, Israel: Interdisciplinary Center Herzliya, June 2014. http://portal.idc.ac.il/he/schools/economics/programs/ugrad/documents/8.pd.

Rudnitzky, Arik, *Arab Citizens of Israel Early in the Twenty-First Century*. Tel Aviv: Institute for National Security Studies, November 2015. http://www.inss.org.il/uploadImages/systemFiles/memo150%20(6).pdf.

Sadeh, Sharon, "Israel's Beleaguered Defense Industry." *Middle East Review of International Affairs*, 5(1) (March 2001). http://meria.idc.ac.il/journal/2001/issue1/jv5n1a5.html#Author.

Sadeh, Shuki, "The Badly Kept Secret of Israel's Trade Throughout the Muslim World." *Ha'aretz*, January 19, 2012. http://www.haaretz.com/israel-news/business/the-badly-kept-secret-of-israel-s-trade-throughout-the-muslim-world-1.408103.

Sahgal, Neha, and Alan Cooperman (eds.), *Israel's Religiously Divided Society*. Washington: Pew Research Center, March 8, 2016. http://www.pewforum.org/files/2016/03/Israel-Survey-Full-Report.pdf.

Saxenian, Annalee, *Regional Advantage: Culture and Competition in Silicon Valley and Route 128*. Cambridge, MA: Harvard University Press, 1994.

Schiffman, Daniel, "The Ideology of Full-Time Religious Study by Married Men in Israeli Ultra-Orthodox Judaism," December 2011. http://papers.ssrn.com/sol3/papers.cfm?abstract_id=1969539.

Schmil, Daniel, "Better Place Asks Court to Appoint Liquidator as Last of Its Cash Runs out." *Ha'aretz*, May 26, 2013. http://www.haaretz.com/business/israeli-electric-car-venture-better-place-seeks-to-dissolve.premium-1.526012.

Schwab, Klaus (ed.), *Global Competitiveness Report, 2010–11*. Geneva: World Economic Forum, 2010. http://www3.weforum.org/docs/WEF_GlobalCompetitivenessReport_2010-11.pdf.

———, *The Global Competitiveness Report, 2016–2017*. Geneva, 2016. http://www3.weforum.org/docs/GCR2016-2017/05FullReport/TheGlobalCompetitivenessReport2016-2017_FINAL.pdf.

Schwartz Altshuler, Tehilla, "Time for the Start-Up Nation to Building a Start-Up Government." Jerusalem: Israel Democracy Institute, March 20, 2017. https://en.idi.org.il/articles/14094.

Schwartz, Dafna, and Ayala Malach-Pines, "High Technology Entrepreneurs Versus Small Business Owners in Israel." *Journal of Entrepreneurship*, 16(1) (2007). http://joe.sagepub.com/content/16/1/1.abstract.

Scott, Mark, "Technology Start-Ups Take Root in Berlin." *The New York Times*, April 29, 2013. https://dealbook.nytimes.com/2013/04/29/technology-start-ups-take-root-in-berlin/.

———, "Corporate Technology Incubators Are Popping Up in Europe." *The International New York Times*, November 25, 2013.

Senor, Dan, and Saul Singer, *Start-Up Nation: The Story of Israel's Economic Miracle*. New York and Boston: Twelve (Hatchette), 2009.

Shalom, Zaki, "The Peace Process: From Oslo Parameters to Unilateral Actions." INSS Insight, No. 267, July 12, 2011. http://www.inss.org.il/index.aspx?id=4538&articleid=2355.

Shwed, Uri, Yossi Shavit, Maisalon Dellashi, and Moran Ofek, *Integration of Arab Israelis and Jews in Schools in Israel, Policy Paper No. 2014.12*. Jerusalem: Taub Center for Social Policy Studies, December 2014. http://taubcenter.org.il/wp-content/files_mf/e2014.12schoolintegration45.pdf.

Silber, Michael, "The Emergence of Ultra-Orthodoxy: The Invention of a Tradition." In Jack Wertheimer (ed.), *The Uses of Tradition: Jewish Continuity on the Modern Era*. New York: Jewish Theological Seminary, 1991. http://www.academia.edu/3260113/The_emergence_of_ultra-orthodoxy_The_invention_of_a_tradition.

Smooha, Sammy, *Still Playing by the Rules: Index of Arab-Jewish Relations in Israel 2012 Findings and Conclusions*. Jerusalem and Haifa: Israel Democracy Institute and Haifa University, 2013. http://www.idi.org.il/media/2519505/Lo%20Shovrim.pdf.

———, *Still Playing by the Rules: Index of Arab-Jewish Relations in Israel 2013 Findings and Conclusions*. Jerusalem and Haifa: Israel Democracy Institute and Haifa University, 2014. https://en.idi.org.il/publications/8731.

Start-Up Ecosystem Report 2012, Startup Genome. San Francisco: Compass Co., October 2012. http://multisite-blog.digital.telefonica.com.s3.amazonaws.com/wp-content/uploads/2013/01/Startup-Eco_14012013.pdf.

Start-Up Ecosystem Report 2015. San Francisco: Compass Co., 2015. https://ec.europa.eu/futurium/en/system/files/ged/the_global_startup_ecosystem_report_2015_v1.2.pdf.

Strauss Group Ltd., 2011, Financial Statement. http://www.strauss-group.com/PageFiles/41425/English.pdf.

Swed, Ori, and John Sibley Butler, "Military Capital in the Israeli Hi-Tech Industry." *Armed Forces & Society*, 41(1): 123–141 (January 2015). http://afs.sagepub.com/content/early/2013/08/22/0095327X13499562.

Teig, Amir, "The Ad Tech Employment Engine: More Than 16,000 Workers in 600 Companies" (Hebrew). *TheMarker*, August 24, 2015.

———, and Inbal Orpaz, "I Have 40 High-Paying Jobs That I Can't Fill" (Hebrew). *TheMarker*, November 7, 2013. http://www.themarker.com/technation/1.2159558.

Tel Aviv University and the Israel Democracy Institute (Ephraim Yaar and Tamar Hermann), The Peace Index, October 2015, November 5, 2015. http://www.peaceindex.org/indexMonthEng.aspx?num=298, http://www.peaceindex.org/files/Peace_Index_Data_October_2015-Eng.pdf.

Tendler, Ilan, "From the Israeli Army Unit 8200 to Silicon Valley." *TechCrunch*, March 20, 2015. http://techcrunch.com/2015/03/20/from-the-8200-to-silicon-valley/.

Timor, Emanuel, and Liran Hason, "Mapping Israel Start-Ups—Automotive 2.0." Vertex Ventures, June 6, 2016. http://www.vertexventures.com/israel/2016/06/mapping-israel-start-ups-automotive-2-0/.

Troen, S. Ilan, "Higher Education in Israel: A Historical Perspective." *Higher Education*, 23(1) (January 1992). https://www.brandeis.edu/israelcenter/about/troen1/HigherEdInIsrael.pdf.

Thompson, Mark, "One Private, Two Armies." *Time*, February, 13, 2012. http://nation.time.com/2012/02/13/one-private-two-armies/.

United Nations E-Government Survey 2016. New York: UN Department of Economic and Social Affairs, 2016. http://workspace.unpan.org/sites/Internet/Documents/UNPAN96407.pdf.

Vale Columbia Center on Sustainable International Investment, Emerging Market Global Players Project, "Israeli Multinationals Back on Track After a Difficult Year," December 12, 2011. http://www.vcc.columbia.edu/files/vale/documents/EMPG-Israel-Report-2011-Final.pdf.

Vital, David, *Zionism: The Formative Years.* Oxford: Oxford University (Clarendon) Press, 1982.

Wadhwa, Vivek, AnnaLee Saxenian, and F. Daniel Siciliano, *Then and Now: America's New Immigrant Entrepreneurs, Part VII.* Kansas City, MO: Kauffman, the Foundation for Entrepreneurship, October 2012. http://www.kauffman.org/~/media/kauffman_org/research%20reports%20and%20covers/2012/10/then_and_now_americas_new_immigrant_entrepreneurs.pdf.

Weiss, Martin A., *Arab League Boycott of Israel.* Washington: Congressional Research Service, December 19, 2013. http://fpc.state.gov/documents/organization/219630.pdf.

Wikipedia, "List of Israeli Companies Formerly Quoted on the Nasdaq." http://en.wikipedia.org/wiki/List_of_Israeli_companies_quoted_on_the_Nasdaq.

World Bank and International Finance Corp., *Doing Business 2012: Doing Business in a More Transparent World.* Washington, DC: World Bank and IFC, 2012. http://www.doingbusiness.org/~/media/FPDKM/Doing%20Business/Documents/Annual-Reports/English/DB12-FullReport.pdf and website http://www.doingbusiness.org/reports/global-reports/doing-business-2012/.

———, Exports of Goods and Services (% of GDP) (database). http://data.worldbank.org/indicator/NE.EXP.GNFS.ZS.

Xavier, Siri Roland, Donna Kelley, Jacqui Kew, Mike Herrington, and Arne Vorderwülbecke, *Global Entrepreneurship Monitor, 2012 Global Report.* March 27, 2015. http://www.gemconsortium.org/report/48545.

Yachin, Dan, and Oren Raviv, *HTIA Annual Review Israel ITC Industry, 2010.* Tel Aviv: Israel High Tech Industry Association, February 2011. http://www.iati.co.il/attachments/272_HTIA%20Annual%20Review%20Israel%20IDC%20Industry,%202010.pdf.

———, *IATI Annual Review: Israel ICT Industry, 2015.* Tel Aviv: Israel Advanced Technologies Industries, January 2015. http://www.iati.co.il/files/files/IATI%20Israeli%20ICT%20Industry%20Review%202015.pdf.

Yadin, Amos, "Five Years Back and Five Years Forward: Israel's Strategic Environment in 2011–2015 and Policy Recommendations for 2016–2020." In Shlomo Brom and Anat Kurz (eds.), *Strategic Survey for Israel 2015–2016.* Tel Aviv: Institute for National Security Studies, 2016. http://www.inss.org.il/publication/strategic-survey-for-israel-2015-2016/.

Yarkoni, Amir, "Why Digital Marketing is a Critical Growth Engine for Israeli Start-Ups." Seperia (blog), October 28, 2013. http://www.seperia.com/

blog/why-digital-marketing-is-a-critical-growth-engine-for-the-new-generation-of-israeli-start-ups/.

Yashiv, Eran, *Arabs in the Labor Market in Israel (Economic Debates, Second Series)* (Hebrew). Jerusalem: Van Leer Jerusalem Institute, January 2012. http://www.vanleer.org.il/sites/files/product-pdf/dispute1_0.pdf.

Yashiv, Eran, and Nitsa Kasir (Kaliner), "Patterns of Labor Force Participation Among Israeli Arabs." *Israel Economic Review*, 9(1): 53–101 (2011). http://www.boi.org.il/deptdata/mehkar/iser/16/iser_3.pdf.

———, *The Labor Market of Israeli Arabs.* Tel Aviv: Tel Aviv University, January 2014. http://www.tau.ac.il/~yashiv/Israeli%20Arabs%20in%20the%20Labor%20Market%20%96%20Policy%20Paper.pdf.

Zatcovetsky, Ilia, and Reuven Gal, *Between Tomorrow (Machar) and Today: The Haredi Academic System at a Crossroads* (Hebrew). Haifa: Samuel Neaman Institute, June 2015. file:///C:/Users/User/Documents/OLD%20COMPUTER/DAVIDS%20COMPUTER/MY%20DOCUMENTS/TechBook/Haredi%20notes/6-434.pdf.

Zicherman, Haim, and Lee Cahaner, *Modern Ultra-Orthodoxy: The Emerging Haredi Middle Class in Israel* (Hebrew). Jerusalem: Israel Democracy Institute, 2012. http://en.idi.org.il/media/2443305/00886012.pdf.

Zinger, Roni, "How Much Does Slaughtering a Kosher Chicken Cost? 1.2 Billion Shekels a Year" (Hebrew). *Calcalist*, January 1, 2016. http://www.calcalist.co.il/local/articles/0,7340,L-3677519,00.html.

Ziv, Amitai, "Virtual Iron Dome: Israeli Cyber Will Save Digital Lives" (Hebrew). *TheMarker*, March 24, 2015. http://www.themarker.com/technation/1.2597804.

———, "Cyber Power: Sales of Israeli Companies 10% of All World Transactions." *TheMarker*, June 25, 2015. http://www.themarker.com/technation/1.2643488.

Databases

Bank of Israel.

International Monetary Fund, World Economic Outlook Database. http://www.imf.org/external/pubs/ft/weo/2012/02/weodata/index.aspx.

Israel Advanced Technology Industries, database. http://www.iati.co.il/category/106/1/accelerators. http://www.iati.co.il/sector/8/incubators.

Israel Central Bureau of Statistics.

Israel Council of Higher Education. http://che.org.il/?page_id=6802.

Organization for Economic Cooperation and Development.

Interviews and Public Addresses (affiliations are those at the time of the interview of event)

Dror Nuhumi, partner, Norwest Venture Partners, January 23, 2012.
Ed Mlavsky, chairman emeritus, Gemini Israel Ventures, January 23, 2012.
Gilad Tuffias, co-founder, TechLoft, January 23, 2012.
Jonathan Goldstein, CEO, Innovo Mimetics, September 10, 2013.
Shlomo Gradman, entrepreneur and chairman, Israeli High Tech CEO Forum, June 26, 2013.
Sigal Widman, strategic and organizational consultant, July 31, 2012.
Zohar Zisapel, Israel Innovation Marathon, Tel Aviv, March 27–28, 2012.

Index

D. Rosenberg, *Israel's Technology Economy*, Middle East in Focus,
https://doi.org/10.1007/978-3-319-76654-6

FSC
www.fsc.org
MIX
Papier | Fördert
gute Waldnutzung
FSC® C083411

Kolibri 360 GmbH
Ferdinand-Jühlke-Straße 7
99095 Erfurt, Deutschland
produktsicherheit@kolibri360.de